Customer Service for Home Builders

Carol Smith

BuilderBooks™
National Association of Home Builders
1201 15th Street, NW
Washington, DC 20005-2800
www.builderbooks.com

Customer Service for Home Builders
Carol Smith

Doris Tennyson	Senior Acquisitions Editor
Jenny Stewart	Assistant Editor
Armen Kojoyian	Cover Designer

BuilderBooks at the National Association of Home Builders

ERIC JOHNSON	Publisher
THERESA MINCH	Executive Editor
DORIS TENNYSON	Senior Acquisitions Editor
JESSICA POPPE	Assistant Editor
JENNY STEWART	Assistant Editor
BRENDA ANDERSON	Director of Fulfillment
GILL WALKER	Marketing Manager
JACQUELINE BARNES	Marketing Manager

GERALD HOWARD	NAHB Executive Vice President and CEO
MARK PURSELL	Executive Vice President Marketing & Sales
GREG FRENCH	Staff Vice President, Publications and Affinity Programs

ISBN 0-86718-562-7

©2003 by BuilderBooks™
of the National Association of Home Builders
of the United States of America

Printed in the United States of America

Cataloging-in-Publication Data available at the Library of Congress
  Smith, Carol, 1946-
    Customer service for home builders / Carol Smith.
      p. cm.
    ISBN 0-86718-562-7
    1. Construction industry–Customer services. I. Title.
    HD9715.A2C87 2003
    690′.8′0688–dc22

2003020956

Disclaimer
This publication is designed to provide accurate and authoritative information in regard to the subject matter covered. It is sold with the understanding that the publisher is not engaged in rendering legal, accounting, or other professional service. If legal advice or other expert assistance is required, the services of a competent professional person should be sought.

–From a Declaration of Principles jointly adopted by a Committee of the American Bar Association and a Committee of Publishers and Associations.

For further information, please contact:
BuilderBooks™
National Association of Home Builders
1201 15th Street, NW
Washington, DC 20005-2800
(800) 223-2665
Check us out online at: www.builderbooks.com

12/03 Armen Kojoyian/SLR/DRC, 2000

Contents

Figures

Carol Smith is the leading customer relations expert for home builders. Her 28 years of frontline experience with customers are immediately apparent in her realistic and practical approach. She has performed over 700 buyer orientations and held the posts of superintendent, custom home sales manager, and vice president of customer relations. In 1999, she founded Customer Relations Professionals (CRP), an international association that provides education and recognition to customer service professionals.

Since 1986 Smith has presented hundreds of educational programs to builders and associates in the United States and abroad, including NAHB's International Builders' Shows, regional conferences, seminars sponsored by the Home Builders Institute (HBI), Custom Builder Symposiums, and Remodelers' Shows. She developed the curriculum for the Home Builders Institute's full-day Customer Service course.

She launched her newsletter, *Home Address,* in 1986 and devotes it exclusively to homebuilding service issues. Smith is an award-winning columnist for *Custom Home* magazine and has written dozens of articles for such publications as *Builders Management Journal, Premier Homes, Building Homes and Profits,* and *Builder* magazine.

She has produced several publications, all published by BuilderBooks, National Association of Home Builders. She has written *Dear Homeowner: A Book of Customer Service Letters; Homeowner Manual: A Template for Builders* (now in its second edition); *Meetings with Clients: A Self-Study Manual for a Builder's Frontline Personnel; Warranty Service for Home Builders,* and customer service brochures.

Introduction

Service is one person doing something for one other person. The person served might be a home buyer, an associate, or another employee. The person serving might be the receptionist, a warranty technician, a superintendent, a selections coordinator, or the company owner. The service might be answering the phone or a question, filling out a form or building a home, providing a repair or conducting a meeting.

Customer reactions to all of these and hundreds of other interactions comprise your company's service reputation. Your goal of course is to have that reputation be a good one so your company can enjoy the typical benefits:

- Referrals are a powerful indication of and reward for customer satisfaction.
- Reputation affects the ease with which you conduct business. From purchasing land to arranging financing, reputation influences the cooperation you receive from the business community.
- Talented people naturally want to work for companies of which they can be proud. Your company's reputation affects employee recruitment and negotiations with vendors and trades.
- Your good reputation can benefit buyers in the long term by increasing the value of their homes.

Do your homeowners mention your company name in newspaper ads when they sell their homes? You do not saunter accidentally into such a service reputation. You need clear service goals and a plan to achieve them. *Customer Service for Home Builders* provides company decision makers with a framework for setting service goals and organizing the activities necessary to reach them. Part 1 outlines fundamental challenges to which every service-oriented company must respond: staff, quality, documentation, internal communication, policies and procedures.

Part 2 examines the chronology of a builder's relationship with home buyers and the issues that builders must anticipate to successfully manage the experience, from aligning expectations through warranty.

Understandably, builders would like to find a simple position to take regarding customer service. The complexity of the new home experience makes working with a one-size-fits-all customer philosophy difficult at best. Sometimes builders perform well; sometimes they create their own problems. Sometimes customers have reasonable expectations; sometimes they do not. Among these extremes every combination is possible.

Often builder service programs are designed to keep the customer under control, enforce boundaries, and fend off attacks. By focusing on their reactive systems, many companies have developed excellent response systems, yet still fail to do as well as they'd like in satisfying their customers. This situation results because such companies invest most of their service resources in responding to problems and they initiate too little.

In today's world, impressing home buyers requires service initiatives. Fortunately, throughout the home buying-building-owning process, you can find hundreds of opportunities to delight and amaze your clients exist. An emphasis on making the entire experience a positive one requires that the traditional roles such as those of sales, selection, construction, orientation, and warranty personnel evolve to a level of sophistication equal to that of the home buyers' expectations.

Initiating

Begin with a commitment to *act as if you like your customer* and these opportunities become visible. This attitude includes appreciation, respect, compassion, tolerance, enthusiasm, and resilience. Initiating service means including proactive elements in your service program. These intentional elements are the positive side of service. By planning an extra attention here or there, you create a context for exceeding customer expectations. These actions are characterized by verbs such as listen, volunteer, provide, anticipate, assist, assure, involve, update, counsel, remind, offer, follow-through, follow-up, invite, confirm, support, and guide.

Responding

Because builders and customers are imperfect, service response systems are still needed along side service initiatives. Companies must sometimes take corrective steps because their performance fails. And they must act quickly to contain the damage when faced with a difficult or dishonest customer.

Within this context builders need a combination of insights, procedures, and trained staff so that their service programs can both initiate and respond. Builders also need the wisdom to recognize when to do which. This wisdom begins with the ability to see a situation clearly, identify pertinent factors, arrive at fair conclusions, and promptly take appropriate actions. The goal of *Customer Service for Home Builders* is to develop that wisdom along side the practical tools.

Your Service Strategy

As you create procedures, establish policies, and train personnel to achieve your customer service goals, you will encounter many questions and make many choices. For some subjects, no universally right method exists. Search for the combination of procedures, and training that results in a service program that is *appropriate* to your company's circumstances. Each organization must find its own way, guided by basic service principles and animated by its own service spirit.

Whether you are refining or updating an existing approach to service or developing one for a new company, you may encounter subjects where you want or need more detail. Because a book of this length is necessarily limited to insights, highlights, and essentials, a Resources List appears at the end. Many of the supplemental sources listed include files on disc, which you can use to tailor materials for your company.

As you proceed, you will define the company's customer service position, create a common vision, and develop a framework for training employees in order to implement that vision. The resulting customer satisfaction will benefit your home buyers and help generate the benefits that a good reputation for service brings.

Chapter 1

Staff

No system or set of procedures is so clever that it is self-propelled. At the heart of any system you will find a person or persons making either a success or a failure out of that system. Little in this book will succeed without the right people–those with the right attitudes, knowledge, and skills. Your objective is to assemble a team that shares your values and goals, then establish policies, procedures, and communication tools and skills necessary to demonstrate the values and achieve the goals.

Walk Your Talk

Values are communicated through actions. The actions of the company owner and management team are profoundly influential. "It starts at the top" applies–the top of the company, the top of the department, the top of the crew. To gain a perspective that explains why wall plaques preaching company values have little impact, consider the multitude of activities that send messages to employees conditioning their attitudes, building their skills, and forming their habits:

- Company reputation, marketing materials, history, accomplishments, awards, and goals.
- Interview process and new hire orientation.
- Employee and operations manuals.
- Daily customer interactions or decisions made by the owner or managers.
- Casual interactions with fellow employees and routine departmental and company meetings.
- Reports, company correspondence, and newsletters.
- Performance appraisals, promotions and firings, and employee recognition or awards programs.
- Physical appearance of the company's offices and work sites.
- Company participation in professional associations and interactions with the surrounding community.
- Social activities, celebrations, and holiday traditions.

The decisions made and actions taken in all of these and other operations areas overwhelm the impact of the most elegant plaques and well-written slogans. Regardless of what you say, what you and other company leaders *do* establishes priorities and sets the tone.

Find and Keep the Right People

Hire people who will be comfortable with your company's values and traditions, priorities and methods. What traits and skills do you need now and over the next few years? Will candidates be able to learn and grow with your company? If you are unable to find the perfect combination of attitude and knowledge, hire people with attitudes that fit your company culture and who have the ability to learn the job. Revising a personality is time consuming and usually unsuccessful.

New Employee Orientation

A thorough orientation helps new employees become effective sooner and provides them with the best chance of long-term success with your company. An effective new employee orientation can also help dislodge the habits an experienced employee brings from a previous employer. Figure 1.1 provides new employee orientation suggestions to help you develop a plan.

Job Descriptions

Well-written and regularly updated, job descriptions provide employees with a sense of direction and boundaries. All job descriptions should include references

---

FIGURE 1.1 New Employee Orientation Suggestions

1. Plan for the new employee orientation to take several weeks, minimally including the entire first day plus regular sessions over the weeks that follow.
2. On the first day:
- Provide an overview and a schedule. Suggest new employee keep a list of questions.
- Schedule times over the next several weeks to check in with the employee for progress updates, feedback, and to answer questions.
- Review the company's history in detail: How and when was it started, significant events, current status, and future vision.
- Verbalize that the company is pleased to have the new employee on board.
- Go over the job description in detail, clearly defining responsibilities and success criteria.
- Review the organizational chart: where does this new employee fit in the organization?
- Address practical details: emergencies, restrooms, kitchen, copier, fax, files, supplies, locks/keys, parking, delivery/pick-up, paycheck procedures and so on.
- Complete paperwork for taxes, insurance, identification, memberships, and so on.
- Tour the office and briefly introduce everyone.
- Visit communities (with a map for notes and landmarks) where homes are under construction.
3. Over the next several weeks, the new employee should spend time in each area. Make the new employee responsible for completing the assigned list of orientation activities within a specified time frame. The Procedure Review Worksheet shown in Figure 1.3 may be useful for this assignment. Alert department heads or individual staff members to expect this visit.
4. Publicize the hiring in a company memo or newsletter, perhaps a press release to the local paper.
5. Consider assigning a mentor to the newcomer for the first six months.

to customer service and continuous improvement. Figure 1.2 offers a job description for a Customer Relations Director. Notice how this employee's relationships to other departments are identified. The NAHB BuilderBooks publication *Job Descriptions* (3rd edition) can offer additional insights.

Performance Reviews

Performance reviews are a means to employee growth and improved performance. Human resources expert Ron Zeedick of HR Strategies offers some guidelines for addressing this important task:

- Set up a time table for performance reviews and follow it faithfully. An effective performance review process has three critical components including a goal planning session for the upcoming year, mid-year progress review, and final year end review. The process will take an experienced supervisor two to three hours per employee per year.
- Reviews should be used as a planning tool for future business and personal development. An employee's goals should tie in directly with the company's overall business plan.
- Include an opportunity for the employee to make observations and participate in setting goals.
- Avoid layering criticism upon criticism. Target one or two areas for improvement and present them in a positive and encouraging tone.
- Back up agreed-upon goals with an action plan. Make goals specific and measurable. If an employee is missing performance targets, call attention to this in daily conversations.

Keep in mind that whom you promote sends a message about the attitudes and behaviors you value. Likewise, when an individual is not performing to expectations, other employees notice and expect you, as a supervisor, to do something about it. Tolerating behaviors that contradict stated company values sends other employees a mixed and confusing message.

Bonuses

In recent years bonus plans have gained popularity in the home building industry. Incentive plans have grown out of the psychological concept known as behaviorism. Behaviorism got its start in the late 1800s. The concept is that people repeat behaviors for which they receive rewards and avoid behaviors for which they receive punishment. Staunch behaviorists believe all human behavior is based on such reinforcements. Opponents argue that people find motivation in their values, ambitions, and commitments–that motivation cannot be imposed from outside.

No doubt the truth is a combination of the two and psychologists will continue to debate this topic. Meanwhile, for home builders, the issue comes down to some practical questions. Can bonus programs produce the intended

FIGURE 1.2 Sample Job Description: Customer Relations Director

Implement [**Builder**] customer philosophy in systems, skills, and attitudes that result in a positive company image and foster repeat and referral customers.

Duties and Responsibilities

Sales & Marketing:
- Orient and update sales personnel regarding service procedures, including preconstruction conference, frame stage tour, homeowner orientation, and warranty service.
- Assist sales staff in responding to buyers' questions about product, procedures, and services.
- Provide sales offices with [**Builder**] Homeowner Manuals and other materials to promote positive and realistic customer expectations regarding [**Builder**] homes and services.

Construction Process:
- Develop procedures for change orders that balance sales and marketing's need for flexibility with construction's need for orderly processes and adequate notice.
- Organize and implement preconstruction conferences for new buyers.
- Work with sales and construction to set site-visit policies.
- Assist sales and construction personnel in resolving home buyer issues during construction.

Delivery:
- Train personnel to perform homeowner orientations in a consistent and positive manner.
- Monitor completion of orientation items, ensuring timely response by construction personnel.
- Develop and maintain a regular reporting system summarizing the number, nature, and completion of orientation items.
- Identify recurring items and work with construction and trades personnel to eliminate them where possible.

Warranty:
- Create procedures for processing routine, emergency, and out-of-warranty items.
- Plan and oversee customer service training for warranty staff.
- Appraise performance of each warranty staff member.
- Maintain a reporting system summarizing the number, nature, and completion of warranty items.
- Identify recurring warranty items and work with construction and trades personnel to eliminate those that can reasonably be eliminated.
- Document and appraise performance of trade contractors in the areas of warranty and customer treatment. Work with trades to improve customer service performance and attitudes.
- Institute after move-in homeowner surveys and focus groups.
- Control costs by efficient use of personnel, appropriate back charges to trades, and feedback to construction and purchasing regarding recurring items.

Supervisory Responsibilities: Warranty Manager, Warranty Administrator, Service Technicians
Reports to: President

results? And if so, under what circumstances? Even those who argue against bonuses acknowledge that incentive plans produce results, at least at first. But positive results are often short term. Ill effects can develop including the following:

- Bonuses become a way of life. Rewards are taken for granted, expected as routinely due for routine work rather than being a stimulus for extra effort and superior results.
- Those who do not "make bonus" become discouraged and may give up. What about those omitted from the bonus system altogether? Customer satisfaction requires teamwork. Leaving some people out can lead to resentment and less than enthusiastic efforts.
- Bonuses may send a message that the job so rewarded has little intrinsic value. The implication is that no one would do the job for its own sake.
- Bonuses can unintentionally reward the wrong behaviors. Often, employees feel manipulated and manipulate back–working the system to "win" against a system they may see as unfair.
- Bonuses may cause company personnel to pull in several directions, as with a program that rewards the vice president of operations for meeting closing quotas, the superintendents for bringing homes in under budget, and the warranty staff for homeowners' satisfaction levels on surveys.

These potential negatives cause real concern. More importantly, bonuses can be a sign that management's response to operational challenges is to throw a bonus program at them and hope for the best. Incentive programs are not intrinsically bad, but their effects can be when bonuses replace effective management. Many companies resort to bonus programs because writing a check is easier than the hard work required to create a functioning team. How can builders who want to use incentives avoid such pitfalls? For bonus programs to make valid contributions to performance while avoiding ill-effects, they should:

- Reflect the company's culture and long term goals.
- Involve participants in their creation.
- Use objectives and criteria that are attainable and still challenging.
- Be equitable–rewarding people for things over which they have control.
- Include rewards that are interesting, relevant, and worthwhile.
- Be specific about duration, preferably short (no more than 6 months per program).
- Include variety–in both objectives and rewards. For ideas, refer to *1001 Ways to Reward Employees* by Bob Nelson.
- Reward individuals or teams based on a comparison of their performance to objective criteria, rather than to other individuals or teams. Everyone can achieve the goals and obtain the reward.
- Include programs for trades.
- Encourage team effort and cooperation.
- Supplement, not replace, respectable pay.

- Exist along side honest communication between management and the front line about any obstacles that prevent employees from doing their best.

Training

Employees can only serve a customer to the extent that they know what to do and how to do it. In a world where change is continuous–sometimes exciting, often unwelcome, and frequently disruptive–training can stabilize, encourage, and challenge companies to new levels of performance. Investments in training pay significant rewards. Typical organizations spend approximately 1.5 percent of revenues on training. Organizations that excel invest as much as five percent. The results of the larger training commitment include positive correlations with higher sales, greater productivity, fewer product defects, less attrition among personnel, and fewer days of work missed.

Employees who are well trained require less time from managers and reflect well on the person who hired them. They make better decisions, solve problems faster, suggest more improvements and innovations, and enjoy their work. In today's job market, where many candidates continuously evaluate career options, training can be one of the factors that leads to retention of valued employees. To grow, whether by increasing share in a current market, expanding into a new market, or venturing into a new product line, companies need employees who are ready for promotion–what sports enthusiasts refer to as "bench strength." Training helps develop that talent.

To gain the greatest benefit from time and money invested in it, training must be well-planned, effectively executed, and followed up by management with coaching activities. The two overriding goals of training are improvement in existing product, processes, and customer treatment; and improvement through innovation–the conception of new product, processes, and service.

Whatever job an individual holds within your company, the more he or she knows about these topics, the more effective that employee can be:

- Communities where your company builds.
- Company operations: policies, procedures, and communication network (meetings and reports).
- Technical aspects of home construction in your region.
- Your company's products and available variations.
- Typical home maintenance tasks and your company's limited warranty guidelines.

Other areas that benefit from attention include professional habits, communication skills, and time management or organizational techniques. Do a training needs assessment, including input from employees. Create a long-range plan including annual training budgets and schedules. Avoid typical vacation months and your traditionally busiest times. Survey available training resources including in-house experts, local HBA programs, local adult education classes, commercial seminars, regional and national conferences, and manufacturers and trades.

Staff meetings offer regular opportunities for training. Thirty to ninety minutes once a month can add considerable knowledge over time. Match the method to the content, audience, purpose, and schedule. Use variety to stimulate interest: demonstration, role plays, "brown bag" discussion over lunch, shared reading, or case studies, for example. When a group has successfully mastered an area of study, a luncheon or other reward is excellent reinforcement. Recognition at staff meetings, certificates of completion, credit toward company certification programs, and so on can be powerful motivators.

Coach

Managers need to stay in touch with what front line personnel face in order to give them support. Employees appreciate management's attention and this level of involvement makes it clear you expect to see change and improvement, thereby preventing the natural inclination to return to familiar habits. Whenever possible, upper management should participate in any formal training you provide. This sends a powerful message to front line employees that the information is valuable and makes effective coaching possible. Coaching activities might include:

- Checking in with employees shortly after training to discuss any concerns.
- Asking positive questions, such as "How would you describe your progress so far?" "What kind of support do you need?" or "Have you encountered any unexpected issues we need to address?"
- Visiting the work site to observe the employee applying the new methods–for example, management visits to listen in on routine client meetings.
- Reviewing key points from training sessions at regular staff meetings, inviting feedback.

Empower

Company owners and managers are sometimes uncomfortable delegating authority to staff. However, being micro-managed frustrates talented people and their having to check with the office slows service to customers. Blindly following procedures no matter what is inappropriate in many situations. When circumstances fall outside the bounds of normal procedures, common sense must find a unique response. The goal is to make decisions as near the front line as possible and as quickly as possible while remaining within the bounds of the company's service philosophy.

The benefits of empowerment–granting authority equal to responsibility–are real and so is the work that must be done before those benefits can be enjoyed. Before you empower your team you must decide "Empowered to do what?" With a clear service strategy supported by well-thought out policies and procedures, management has a framework for empowerment. As they see well-trained employees exhibit good judgment, managers become more comfortable giving greater authority.

Cross Train

Friction among sales, construction, and warranty–"the natural enemy syndrome"–usually signals a lack of understanding of each other's goals, methods, and challenges. To reduce this, cross train employees. Benefits include greater efficiency, flexibility, and co-operation. Understanding and mutual respect translate into supportive daily behaviors such as the superintendent reminding a trade to complete warranty items or the warranty manager meeting with a prospect to answer technical warranty questions in support of a sales person. The Procedure Review Worksheet in Figure 1.3 offers a simple method employees can use to document an organized study of company procedures outside their area.

Trade Contractor Service

Homeowners apply the same criteria to a trade's performance as they apply to the builder's staff: response time, attitude, and quality of the work. In hiring trades, few builders would leave specifications, scheduling, and prices to chance. Likewise, the impact of trades' service on customer satisfaction is too important to leave to chance. Adding customer service to early discussions with potential trades raises service to the same level of importance.

In response to this need for increased communication, some companies designate a staff member as the liaison between the trades and the company. Although the position was conceived with large companies in mind, many of the specific duties are applicable to firms of any size:

- Meet with each trade or supplier and conduct a thorough orientation into company processes.
- Work with construction and trades to develop written quality checklists based on scopes of work, field experience, and feedback from homeowners and warranty.
- Serve as chairperson for a trade contractor alliance, facilitating monthly meetings to identify improvement opportunities. Follow through on approved suggestions.
- Review work in the field with superintendents to confirm that specifications and scopes of work are met. Recommend contract revisions to the purchasing agent as needed.
- On a quarterly basis, survey field personnel and warranty staff for evaluation of trades and suppliers. Survey trades and suppliers for their comments on company personnel and policies. Summarize the results.
- Organize annual social events and recognition, including a summer picnic and a winter party to celebrate successes and show appreciation of trades and suppliers.
- Notify sales and design center staff of changes in production.
- Maintain sales office binders of consumer product warranties for Magnuson-Moss Act compliance.

FIGURE 1.3 Procedure Review Worksheet

Step 1: Documentation

Gather copies of the documents related to this procedure, list them below, and check off each as you read it.

_____

_____

_____

_____

_____

_____

_____

Step 2: Interviews

Interview one or more individuals who perform this procedure. Attach notes as needed.

Date   Name

_____   _____

_____   _____

_____   _____

_____   _____

Step 3: Observations

Observe one or more appointments with clients to see examples of this procedure being performed. Attach notes as needed.

Date   Name

_____   _____

_____   _____

_____   _____

Part 4: Conclusions

Record your thoughts and conclusions regarding this procedure-what are the strengths and where are the opportunities to improve this part of the new home process? Attach your notes following this sheet.

- Meet monthly with warranty to review trade responsiveness to work orders and customer treatment. Identify recurring items and coordinate with construction and trades to eliminate them.

Service Orientation

The contract between the company and the trade contractor should reference your service policies and document the trade's warranty obligation in particular. You can find detailed information and sample wording in BuilderBooks' publications titled *Contracts with the Trades: Scope of Work Models for Home Builders, Contracts and Liability, and Form Builder: Contracts.* But signing such a contract is not the end of defining your service expectations.

Whether your company personnel roster includes a trade contractor liaison or not, when you hire a new trade, set an appointment to meet with that firm's owner and the trade's warranty service person. Create a standard packet with copies of all applicable forms for these meetings to assure consistency. Add to this a copy of the homeowner manual information that relates to the trade's work. Time spent covering this information is an investment in your reputation.

A sample trade contractor service orientation agenda appears in Figure 1.4. Include points about customer service throughout the relationship, not just during warranty. Your customized version of the Warranty Repair Guidelines (see Fig. 2.3) will provide an excellent base for discussion of daily warranty service behaviors.

Difficult Trades

If a trade contractor is out of step with the rest of the team, you have three options: (1) solve the problems with the trade, (2) replace the trade, or (3) continue putting up with unsatisfactory service. While this last choice is unacceptable to a company concerned about its service reputation, turnover of trade contractors is time-consuming and causes confusion in homeowner services.

A sincere effort to resolve the situation is worth a try. Schedule a problem-solving session with the trade company owner and the service person. Have specific examples of unacceptable service events so the trade can learn what needs to change. Approach the discussion in a positive manner–"Perhaps we could have communicated our service expectations more clearly." Give the trade an opportunity to discuss procedures that impede good service. Watch for improvement. If two attempts fail to produce improvement, replacing the trade may be your only choice.

The decision to replace a trade should reflect a balanced evaluation. How is the trade's performance in other areas? How many homes are involved? Changing trades for the last two houses in a subdivision of 116 homes makes little sense. However, keep in mind that if word gets around among other trades that your company tolerates poor service performance, other trades may put forth less effort. This dilemma is best prevented. As with company employees,

FIGURE 1.4 Trade Contractor Customer Service Orientation Agenda

Prepare standardized trade contractor orientation packets including the meeting agenda and copies of relevant literature, forms, and reports.

Pre-sale Services:
- Review plans and specifications, suggest improvements where appropriate.
- Review and confirm applicable maintenance and warranty information in the homeowner manual.
- Price optional items.

During Construction:
- Adhere to all applicable safety practices.
- Follow quality management system.
- Clean up scraps daily.
- Price change orders promptly.
- When buyers visit:
  - Turn radio down or off.
  - Avoid foul or vulgar language.
  - Avoid negative comments about personnel, specifications, or the work of others.
  - Refer buyers to sales person for answers to questions.

Move-In/Transition Service:
- Establish a rapid response procedure for orientation items.
- As needed, schedule system start up and client demonstration (applicable to home theater, pool, sprinkler, and security companies).

Warranty Coverage:
- Trade contract clause regarding warranty
- Start date and duration of coverage (including the "grace period")
- Limited warranty document
- Purchase agreement clause regarding warranty

Warranty Procedures and Paperwork:
- Standard warranty contacts
- Warranty service request form
- Inspection form
- Work order
- Confirming work order for emergencies, if applicable
- Pending work report
- Repercussions of late work orders
- Paid service work
- Back charges
- Recurring items

hire carefully, orient thoroughly, and communicate regularly to create a service-oriented team.

Technology: Electronic Staff

Nearly every technique in this book can be supported in some way by technology. Overhead can be kept to a minimum, stress reduced, records kept current and comprehensive, timely and accurate reports generated, and communication among all parties expedited with appropriate technology. For example, networking records of selections and change orders can prevent delays and costly errors, allow superintendents to answer trade questions faster, and increase buyer confidence. Option pricing is easier to update when pre-priced option information is stored in a database linked to each sales office. Computer generated warranty reports are more likely to catch recurring items. Think of technology as an electronic member of your staff, a partner who helps all personnel serve your customers more effectively.

Chapter 2

Quality

The concept of quality cannot be separated from customer satisfaction. Both begin with the customer's expectations, and the customer decides how well the builder met those expectations.

"This is our third new house. We haven't had serious problems with any of our homes, but none of the builders have been spectacular. Things go wrong; they fix them. Sometimes you have to call three or four times, sometimes you have to yell. Sometimes when they fix something, they fix it okay, but they don't always fix it right. It's as though they do it just to shut you up, not because they should."

These remarks came from a working mother of two who loved her Midwest home. She believed her builder had integrity and the trade contractors were courteous. She wonders why receiving the quality she paid for had to be such a battle. Today's better informed buyers want some answers or they want some changes. If it is possible to build one straight wall, why can't they all be straight? If dips and humps in floors are normal, why don't all floors have them? If it is possible to paint one wall correctly, why can't they all be painted correctly?

Strengthening your recovery system to provide good repairs faster is better than ignoring the situation but does little to eliminate defects. An essential parallel response is to build with greater precision and pursue continuous improvement. To do so requires a quality management system.

Admittedly, standards are a moving target, a tug of war between quality and price, scheduling and skills. A standard can be as clear and concise as a measurement on a ruler or as subjective as a homeowner's emotional opinion of a carpet seam. The standards of each of us change with time and new knowledge. Some are matters of personal taste, style, or even nostalgia. Builders struggle to assemble a set of standards that will appeal to their target market and struggle further to build homes that consistently meets those standards at prices buyers are willing to pay.

Without a quality management system everyone who contributes to the home is operating on his or her own definition of quality instead of what the company decides. The results will be correspondingly varied and unpredictable. This unpredictability creates conflict with customers when the salesperson promises them one thing and construction delivers something different. On the other hand, detailed, written company standards:

- Provide a framework for training personnel, hiring trades, and managing daily work.

- Form the basis of representations the sales people make to customers about quality.
- Create a foundation for warranty tolerances.

Responding to the "Yeah, Buts"

Mention "quality management" and most builders quickly acknowledge its potential contribution to customer satisfaction. However, just as quickly, the "yeah, buts" start.

"Yeah, but my people are all overworked as it is." Yes, overworked with rework, complaining buyers, documenting back charges, and replacing trades who quit or got fired.

"Yeah, but I can't afford to add anyone else to our payroll." Perhaps because you spend so much on repairs and concessions in the sales office to offset a questionable reputation? In any case, quality management can be implemented with current staff in most companies. Once an effective quality management system takes hold, warranty costs begin to drop. Some builders have seen such costs reduced by up to half.

"Yeah, but we have too much paperwork already." The paperwork that supports quality management reduces conflict with customers, warranty work, and the paperwork related to both.

"Yeah, but we're not trying to build perfect houses." Quality management targets predictability and consistency rather than perfection. You set the standards; quality has the meaning you specify. The end product is what you intended, not just what happened.

"Yeah, but we expect the trades to do the work right, that's what they get paid for." Anyone can make a mistake. Besides, how do you know you have communicated clearly? One of your obligations to customers is to confirm that materials you promised are installed correctly. One of your obligations to your company is to confirm you get what you pay for.

"Yeah, but our supers are experienced professionals. They know what's right. They don't need a written system." Experienced professionals can have a bad day or get interrupted. Besides, memories fade quickly and can neither be put in the file nor tracked objectively.

"Yeah, but we already have a quality inspection. It's just before delivery. We make a list and fix anything we find. There's no need to slow everyone down with more steps." Home buyers believe otherwise. When errors are overlooked, good work added on top of them is often torn apart later to correct the original problem. Errors that accumulate until near delivery make buyers nervous and reduce their confidence in the quality of their home. Buyers familiar with quality management principles from their own work suffer even more frustration.

"Yeah, but I read an article on quality management that said it takes two years to get that kind of system in place. I don't have time to wait for something like that." Where will you be in two years if you don't get started? Quality management is an attitude and a process; both can contribute to the successful operation of your business and higher customer satisfaction.

The traditional seat-of-the-pants approach to quality has had ample opportunity to prove itself in the home building industry. It works inconsistently at best. As many companies have already discovered, replacing the seat-of-the-pants approach with a planned, written quality management system produces the desired results.

Steps to Quality Management

Quality management can be complex-loaded with statistical process formulas, complex charts and graphs, endless meetings and reports–or it can be simple: set and hit your quality goals. A simple system appeals to most field professionals.

Put It in Writing

To serve everyone effectively, your company's standards must be accepted by all the participants:

- Trade contractors who are expected to meet the standards
- On-site supervisors who are expected to enforce them
- Sales people who are expected to present them to customers
- Customers who are expected to live with them
- Warranty staff who are expected to keep the product within them

Aligning all these stakeholders requires putting your standards in writing. This is one of the most time consuming yet important tasks in establishing your quality management system. Define, in terms not open to interpretation, exactly what standards are to be achieved at each phase of construction. If you are just getting started, framing is one of the best places to begin. If a home is framed badly you cannot put enough caulk and paint on it to make it look good. Framing errors are costly to repair after all the finish work is in place–not to mention the negative impact on your home buyers' goodwill.

Besides framing, documented quality confirmations are conceivable at many points. Figure 2.1 suggests a few to get you started. BuilderBooks publications such as *Production Checklist for Builders and Superintendents* and *Building Quality: An Operations Manual for Home Builders* offer more examples and guidance.

Consider input from many sources as you set your standards. Involve those who will be using the standards in creating them. A team approach is needed for this to work effectively, with all parties comfortable with the goals and procedures. Without this involvement, the participants may later go through the motions but will lack the spirit of the system. The discussions that lead to written quality standards clarify and unify company thinking about quality. Select and list the points to be confirmed concisely and allow space for notes

FIGURE 2.1 Potential Topics for Construction Quality Confirmation Checklists

Begin your quality confirmation checklists with these areas, adding others as needed:

- ☐ Footers or caissons
- ☐ Foundation
- ☐ Waterproof
- ☐ Backfill
- ☐ Rough frame
- ☐ Roof
- ☐ Exterior trim
- ☐ Insulation
- ☐ Drywall
- ☐ Prepaint
- ☐ Interior finish
- ☐ Exterior finish
- ☐ Final grade
- ☐ Landscape

about corrections. These lists evolve over time based on new methods and materials as well as feedback from everyone involved.

Depending on the phase being inspected, these quality confirmation checklists might contain from just a few to several hundred points. As an added benefit, the materials you are creating will make excellent training tools for new employees and trades. Figure 2.2 shows an example.

Communicate

With in-house staff, regular meetings offer a forum for discussing quality standards and related issues. Do not stop there; get out in the field and walk houses. Study quality (yours and your competition's) at each stage of construction with your employees. Look closely and talk about what you see.

For trades, communication about quality begins with the purchasing manager—first in conversation and later in a written contract. Select your trade contractors based on balanced criteria that include their commitment to participate in your quality management efforts. An attachment to trade contracts details the required standards, referencing applicable codes and adding your particular quality standards. Usually called scopes of work, these documents reduce misunderstandings and increase consistency. Examples of scopes of work can be found in Builder Books' publication, *Contracts with the Trades* by John Fredley and John Schaufelberger.

FIGURE 2.2 Rough Frame Quality Confirmation

Job # _____   Date _____

	Acceptable	Correction Needed
1. Plates properly located on sill		
2. Plates properly grouted		
3. Framing complies with plans/code		
4. Sway braces properly installed		
5. Trimmers under headers flush w/ studs and shimmed		
6. Nuts installed on sill bolts		
7. Thermoply properly installed		
8. Tele posts plumb and bearing on pads		
9. Stairs back nailed		
10. Studs/walls straight & plumb		
11. Joist hangers, truss clips installed		
12. Bathtubs shimmed		
13. Attic access framed		
14. Heat boots covered		
15. Floor joists quality, placement		
16. No unnecessary holes in Thermoply		
17. Thermoply placement avoids infiltration		
18. Pipes protected from nail damage		
19. Duct work and return air per plan		
20. Duct work suspended to eliminate vibration		
21. Gas line properly installed		
22. Heat runs insulated-outside walls		
23. Interior cleaned of trash		
24. Inspection card signed		

By _____

Monitor the Work

As the homes are built, trades and superintendents should pause regularly to look at them. Do they meet the standards intended for that phase of construction? Return to the question, "At this point what should this house look like?" With a written quality confirmation form, needed corrections are documented and then checked off when they have been completed. The random effects of human memory, personal opinions, and perhaps even mood are dramatically reduced.

This quality confirmation function can be fulfilled by a full-time inspector or by existing personnel "trading places" to check each other's work. Some companies ask each trade contractor to confirm that his or her work meets the required standards and turn in the completed checklist to the superintendent. The superintendent makes a judgment call as to whether to recheck that work. Based on track record, some trades' work is spot checked by the superintendent, others are inspected every time until they meet desired quality levels consistently.

Correct or Adjust

When you find errors at the earliest possible time they can be corrected for the least possible cost, in dollars and customer goodwill. Consider all aspects of the process when corrections are needed:

- Redesign some detail of the product.
- Select different materials or methods.
- Increase training for company personnel or trades.
- Communicate more clearly with trades.
- Revise selection criteria for hiring trades.
- Adjust workload.
- Change schedule or sequence.
- Adjust customers' expectations.

Cleaning and Quality

A builder received the following comments in a letter from an unhappy homeowner: "During construction, I asked if the furnace would be cleaned of construction debris and was told that it would be taken care of. Today I opened the furnace to remove and clean the filter and found the compartment full of sawdust and loose screws. After I removed the filter, I found a Ruffles Potato Chip bag in the compartment next to the filter. In the process of removing the potato chip bag, I cut my finger. I used my shop-vac to clean up the mess in the furnace. Based on this and other incidents that occurred during the construction of my home, I have developed a very low opinion of your organization. Communications between you and your contractors is terrible and your quality control is worthless. I go out of my way to tell my friends and co-workers what a bunch of bumbling screw-ups you are. The saddest part of all is that the conduct of your company is typical of the construction industry."

Construction sites are notoriously messy. See this as an opportunity to beat the competition and impress home buyers. Expect trades to remove what they bring in—such as cartons—and clean up the debris created by their own work (and lunches). Once they become accustomed to this responsibility, most trades admit they prefer it. The cost is small compared to the returns:

- Buyers' confidence that their home is well built soars when they observe this level of attention to detail and control by the company.
- Tradespeople take more pride in their work and show more respect for the work of others.
- Quality confirmations are easier to perform in a clean home.
- Showing clean homes to prospects is a pleasure for your sales team.
- Keeping track of delivered materials is easier, reducing theft and damage.
- Homeowners in occupied homes near-by appreciate that construction trash is not blowing all over their yards.

Warranty Feedback

The insight that the warranty department can offer is too valuable for builders to ignore. Listening to feedback from warranty can spiral a company to better quality homes, greater customer satisfaction, and lower warranty costs. But as you analyze recurring items, you may find more opportunities for improvement than you bargained for. You'll need a method of prioritizing to select appropriate targets for elimination.

- How frequently does the problem occur? Three percent of your homes? Thirty-seven percent?
- What are the repair costs in terms of administrative time and dollars?
- What is the hassle factor for the homeowners? (Number of appointments, dust, noise, odor, and inconvenience?)
- What is the cost of prevention?

For example, if an exterior repair item shows up in seven percent of homes, takes one short visit and $16 to repair, but costs $48 to eliminate up front, you would probably continue to repair this item. However, if an item occurs in 41 percent of your homes, is a serious nuisance to homeowners, and costs $187 to repair, but costs $62 to prevent up front, it is worth making changes to prevent.

Once a change is approved, the next step is to implement it fully. Ask "Who else needs to know?" Check all possibilities including purchasing, salespeople, accounting, draftspeople, design center staff, trade contractors, superintendents, show home displays, contract documents, and homeowner manual. Consider whether buyers under contract need to be told of the change.

Process Quality

You can apply a quality management mind set to processes also by setting standards for each part of the home buying experience. As an example, envision the ideal closing. List standards for each aspect of that process. You might begin with something as simple as whether the customers can find the closing location and arrive on time. Are buyers prepared with correct funds, utilities transfer orders placed, evidence of insurance, and any mortgage contingencies satisfied? Does the closing agent greet the customers warmly and promptly begin the process, presenting documents in a logical order? Now compare the actual events with your standards and note where variations occur, looking for opportunities to improve. For instance, if customers frequently arrive late because the closing office is difficult to find, you could provide a map.

Face to Face with Customers

Company personnel and the 40 to 50 trade contractors you hire will each have different attitudes and skills for working with customers. If you hire someone who says that he or she provides exceptional service, your response may be "Great, we insist all our homeowners receive exceptional service." The problem comes from not checking out how that person's definition of exceptional service aligns with yours. Again, quality management methods offer support. Identify your standards, communicate them, monitor the results, and make changes to meet your goals. For example, suggested guidelines for conducting warranty repair visits appear in Figure 2.3. As with your other quality standards, you will gain cooperation from your team by inviting them to help develop these standards. Contributing to the process creates ownership and enthusiasm, both of which improve results.

FIGURE 2.3 Warranty Repair Visit Guidelines

Preliminaries

1. Check the work order to see if you have any questions, if so, contact the warranty administrator for help.

2. If the work order includes a "work date" make note of it on your calendar.

3. Confirm that you have all necessary tools and materials, or order required items.

4. Estimate the amount of time the work will take and arrange your schedule to complete the repair in one trip as often as possible.

5. If no "work date" appears on the work order, contact the homeowner to schedule a repair appointment.

   • Give yourself a 30-minute range to allow for traffic and other unexpected events.

   • Minimally commit to an a.m. or p.m. time frame.

6. If you will be late, call the homeowner as soon as you know, apologize and offer to re-schedule.

Upon Arrival

7. Park in the street.

8. Refrain from smoking on the homeowner's property.

(Continued)

FIGURE 2.3 Warranty Repair Visit Guidelines (Continued)

9. Take the work order and a business card with you to the door.

10. If you have not met the homeowner before, introduce yourself and give him or her a business card.

11. Avoid using first names until the homeowner invites you to do so.

12. Remove or cover your shoes before entering the home.

13. Review the repair item(s) and their location(s) with the homeowner if necessary.

14. If the homeowner changes or misses the appointment, note this on the work order and inform the warranty administrator.

15. If the homeowner is not home when you arrive:

    - Wait a minimum of ten minutes and then leave a door hanger.
    - Notify the warranty administrator.
    - Contact the homeowner to re-schedule.

16. While an appointment may not be necessary, contact the homeowner prior to performing any exterior work. Leave a door hanger after completing exterior work so the homeowner knows you were there.

Repair Work

17. Check the work area. If any cosmetic damage exists, note that and ask the homeowner to initial your note.

18. Be aware of your surroundings to ensure that children, pets, and the homeowner's belongs will not be harmed.

19. Protect counters, floors, and other surfaces from tools, tool boxes, drips, and dust with a drop cloth or other material.

20. Ask the homeowners to remove personal items or furniture from the work area.

21. If conditions are inappropriate for the work you need to do, leave and reschedule the repair appointment.

22. If you need to use the homeowner's electricity or water to perform the repair, ask permission first. Avoid using the homeowner's phone or bathroom unless they volunteer that you are welcome to do so.

23. Perform the needed repair(s) efficiently and within [Builder] standards.

24. If you are uncertain about how something should be done, look at the show home or contact the Warranty Manager.

25. If a follow-up visit is necessary to complete the work, schedule it immediately and inform the warranty administrator.

26. Refrain from making any negative comments to homeowners about the work of others, personnel, the company, or the community.

27. If you notice a serious problem, report it to the warranty office immediately.

28. Use the "10-minute rule" to screen new items. Note any new item you repair on the work order so the warranty file is complete.

(Continued)

FIGURE 2.3 Warranty Repair Visit Guidelines (Continued)

29. Immediately report any damage caused by tool boxes, dirty shoes, mistakes, or accidents so appropriate corrections can be arranged.

Conclusion

30. Clean-up the work area: remove all tools, materials, dust, and debris generated by the repair.

31. Make brief notes describing what you did to repair the item(s).

32. Inform the homeowner that you are finished.

33. Request the homeowner's signature on the work order. If homeowner is unavailable or prefers not to sign the work order, make a note of that.

34. Sign and date the work order.

35. Turn the work order in to the warranty office.

Chapter 3

Forms and Documents

While some dream of paperless systems, most of us wallow in the stuff. This can be a good thing if it is the right paper and you use it well. Make *put it in writing* the mantra of your organization and supply well-planned support for that goal. This chapter offers suggestions for documents customers receive as well as several generic forms that might be used by personnel at any point in the builder-buyer relationship. Additional forms appear in later chapters where specific procedures are discussed.

Too Much of a Good Thing

Builders share procedures and forms on every aspect of building—including customer service. At conventions, local meetings, and casual gatherings, builders exchange materials as never before. This cross pollination has helped many companies improve service. However, it does increase the volume of paper builders provide to home buyers.

Customers stimulate increases in the mountain of information. For instance, after a confrontation with a homeowner about landscaping, a builder is likely to add several more pages of landscaping explanations and limitations. An orientation rep might create a handout regarding orientation procedures. Trade contractors frustrated with work orders about some aspect of their work create a fact sheet—about brass fixtures, shrubs, cabinets, or so on. Now another item is added to the stack customers receive—and, another font, format, paper quality, and writing style.

Seldom does anyone ask why what already exists fails to produce the intended result. Everyone reasons that because buyers show no sign of being aware of policies and even less inclination to follow them—they obviously need to be told again. New handouts overlap the originals—as if twice of much of what did not work the first time will solve the problem.

As mismatched and overlapping information accumulates, each new version often delivers a slightly different message. This meandering opens the door to multiple interpretations by customers and staff. One employee knows what one page says, another staff member goes by an older or newer version. Expecting personnel to speak with one voice is unrealistic with so many scripts in circulation.

Recognizing the massive nature of their materials, some builders withhold information, dribbling it out throughout the process. The dribbling approach (for instance, delivering warranty information at the orientation or closing) recognizes the symptom while ignoring the cause. This method also wastes the best opportunity to align home buyer expectations: during the early stages when buyers are in an informational gathering mode and are most open to new data. Buyers can also say, "You didn't tell me this when I signed the contract, do I don't have to agree to it." A better approach is to take charge of the materials.

Take Charge

When builders organize details so that buyers can make sense of them comprehension, cooperation, and satisfaction increase. To streamline information, consider the following steps:

- Collect one copy of each document, form, brochure, or handout buyers receive.
- Assemble them in the order your customers typically encounter them.
- Read all of them, watching for contradictions and repetition. As you do this, circle every no, not, or similarly negative word or phrase for possible revision.
- Keep the end users in mind. How many places do homeowners need to look to find all of your standards and policies on any one subject (change orders, plumbing, landscaping)? When information is organized employees can make more effective use of it as well.
- Where appropriate, reorganize. Keep like things together and unlike things apart. Say what you have to say clearly, one time.
- Once your material is updated, coordinated, and organized, commit to keeping it that way. Yearly revisions blend valid new information into the existing materials.

New Materials

There are a lot of things you can do. The challenge is to identify what you should do—and then do that well. Any new material, form, or format must be thoughtfully compared to existing materials and information. Evaluate new materials or ideas systematically.

- Does the new information or format fit your company culture and support your desired image?
- Does this new presentation serve a need not addressed with current materials?
- Would it be better to fine tune the existing information or should you revamp completely?

For years builders gave too little information to their buyers. Now many are giving too much; everything they created and anything else they find as well. Appropriate content and effective use, not volume, will do the job. Buyers welcome correct, complete, well-organized information as a sign that their builder understands the magnitude of the task and can successfully manage a myriad of detail.

Friendly Tone

Wording can be clear and still friendly, making the boundaries easier to accept from the buyers' perspective and the information easier to communicate from yours. Read through every printed item you give to customers with a red pen in hand. Circle every "no," "not," or variation of the two. If your paperwork is like most companies, you will find more than you expected. Next, work to re-word the information to convey the same boundary and still sound hospitable. Can you use another word? Consider words such as avoid, prohibit, exclude, or unavailable. Or ask "What is true?" and describe that instead of what is not true. For example:

- Replace "Do *not* expect a dust free surface on your hardwood floors." with "A dust free surface is *impossible* to attain when finishing hardwood floors."
- Replace "Our warranty does *not* cover concrete cracks." with "Our warranty *excludes* concrete cracks."
- Instead of "You *cannot* make changes at the preconstruction conference" say "*We will schedule* your preconstruction conference after you have made the last change to your new home plans."
- If the client asks for a change at the preconstruction conference in spite of this preparation, avoid saying "We *can't* make changes at this point" by saying "The time to make changes *has passed.*"
- Instead of "Our warranty does *not* provide any attention to this item" say "This *is* a homeowner maintenance item."

Books Judged by Their Covers

Spotted, crooked, blotched, smeared, and generally unreadable, poor-quality paperwork sends the wrong message to customers. Fourth generation copies of copies with the inevitable liver spots, mismatched margins, and slanted text are unacceptable. Such materials make buyers wonder, "If this sloppy paperwork is acceptable to the builder, how good can my house be?"

Consider the typical colored brochures with which builders woo prospects during the sales process. After sparing no expense to attract buyers' attention and obtain a sale, builders too often miss the opportunity to continue sending a message that conveys high quality throughout the rest of the process. Putting information in writing helps create legitimacy and credibility. However, much of the authoritative image is sacrificed if those materials are of poor quality. The customer satisfaction that produces referrals and a terrific reputation comes from the professional handling of every last detail.

Examine your materials. Take a look at your homeowner manual. If you are still using a $3 white binder and inserting a title page into the clear vinyl pocket at the front, if you have an administrator who types section titles and inserts them into old-fashioned index tabs, consider taking a different approach. An update may be in order to ensure that all items customers see reinforce your quality image. Strive for a reasonable uniformity in document appearance to demonstrate control over the myriad of details. Establish standards for company paperwork. Begin your quality list with points such as these:

- Logo appears in the same location on each form.
- Builder's address and contact information is current.
- The title of the form is clear.
- Details progress in a logical order.
- Wording is positive and friendly; where possible no, not and other negatives are avoided.
- Adequate space is provided where information needs to be filled in.
- Spelling, grammar, and punctuation are correct.
- Fine print is avoided (if policies are embarrassing, perhaps they need to be changed).
- ALL CAPS are avoided (they are the typographical equivalent of SHOUTING).
- No carbon required (NCR) forms are used where appropriate.

Before creating a new form, check whether an existing form can be modified to include the new material. When replacing an outdated form, collect and dispose of all copies of the old version.

Homeowner Manual

A comprehensive manual, delivered at contract, informs buyers about normal events of the building process, offers a convenient place to organize materials that accumulate, and serves as a reference after move-in. Strive to channel all information you want buyers to have through your homeowner manual. Suggestions about use of this manual appear in Chapter 6, Aligning Customer Expectations.

Bulleted Summaries

Divide your manual into sections with professionally printed tabs and add a bulleted summary of the key points as the first item for each section. This allows sales people to provide an overview of the contents in just a few minutes. Possible bullets for a "Financing Your Home" section, reprinted from Builder Books' *Homeowner Manual Template*, appear in Figure 3.1 to illustrate.

Alphabetical Maintenance Hints and Warranty Guidelines

Builders struggle to make the separation of maintenance and warranty clear to buyers. A significant aid to that effort is appropriate formatting of this information. By thinking about the end users—in this case the homeowners and the builder's warranty staff—a couple of key points become evident. First, informa-

---

FIGURE 3.1 Bulleted Summary of the "Financing Your Home" Section of a Homeowner Manual

- Loan Application Checklist-lists the documents and information typically needed to complete the loan application form.
- Loan Application Paperwork-an overview of the forms involved in processing your application.
- Underwriting-key points to be aware of regarding the loan approval process, take special note of contingencies that may apply.
- Loan Lock-lock your loan only after [Builder] has provided you with a written delivery date confirmation.
- Loan Closing-avoid changes to your financial circumstances to protect your loan approval.
- Down Payment Worksheet-to assist you in determining the amount you have available for your down payment.
- Monthly Payment Worksheet—to assist you in estimating the monthly payment amount for your new home mortgage.

---

tion must be easy to retrieve. Homeowners are unlikely to search for answers very long; they will call the warranty office instead. When your personnel need to show details to homeowners, they want to appear confident and professional–in other words, they do not want to search long either.

Traditional methods of formatting warranty standards fail in this regard. Matrices including obscure headings that mean little to most homeowners are almost completely ignored and staff can seldom find what they need in them. Likewise, repetition of the word "Defect" or "Deficiency" in bold print is of little help in locating the subject needed and may make normal occurrences appear to be negatives: If shrinkage and cracking of caulking is to be expected, why list this behavior as a "defect"?

An alphabetical listing of the home's components (air conditioning, brass, bricks, cabinets, and so on), supported with appropriate sub-topics–also in alphabetical order–creates a more useful reference. By arranging the information under each component in two categories, with "Homeowner Maintenance Guidelines" first and "[Builder] Limited Warranty Guidelines" next, the separation between the two becomes more clear. Figure 3.2 shows an example from Builder Books' *Homeowner Manual Template*. Small volume custom builders can print each homeowner manual separately, using the search and replace command to insert the homeowners' names at each list of maintenance guidelines ("Tom and Mary Jones' Maintenance Guidelines"), making the distinction even more clear.

Limited Warranty Agreement

Before closing on the company's first home, a builder should have a warranty document, warranty guidelines, and a plan for warranty service in place. Addi-

FIGURE 3.2 Sample Homeowner Manual Entry: Asphalt

Homeowner Use and Maintenance Guidelines

Asphalt is a flexible and specialized surface. Like any other surface in your home, it requires protection from things that can damage it. Over time, the effects of weather and earth movement will cause minor settling and cracking of asphalt. These are normal reactions to the elements and do not constitute improperly installed asphalt or defective material. Avoid using your driveway for one week after it is installed. Keep people, bicycles, lawn mowers, and other traffic off of it.

Chemical Spills. Asphalt is a petroleum product. Gasoline, oil, turpentine, and other solvents or petroleum products can dissolve or damage the surface. Wash such spills with soap and water immediately, and then rinse them thoroughly with plain water.

Hot Weather. Avoid any concentrated or prolonged loads on your asphalt, particularly in hot weather. High-heeled shoes, motorcycle or bicycle kickstands, trailers, or even cars left in the same spot for long periods can create depressions or punctures in asphalt.

Nonresidential Traffic. Prohibit commercial or other extremely heavy vehicles such as moving vans or other large delivery trucks from pulling onto your driveway. We design and install asphalt drives for conventional residential vehicle use only: family cars, vans, light trucks, bicycles, and so on.

Sealcoating. Exposure to sunlight and other weather conditions will fade your driveway, allowing the surface gravel material to be more visible. This is a normal condition and not a material or structural problem. You do not need to treat the surface of your asphalt driveway. However, if you choose to treat it, wait a minimum of 12 months and use a dilute asphalt emulsion, rather than the more common coal tar sealant. Hairline cracks will usually be filled by the sealing process. Larger cracks can be filled or patched with a sand and sealer mixture prior to resealing.

[Builder] Limited Warranty Guidelines

We perform any asphalt repairs by overlay patching. [Builder] is not responsible for the inevitable differences in color between the patch and the original surface. Sealcoating can eliminate this cosmetic condition and is your responsibility.

Alligator Cracking. If cracking that resembles the skin of an alligator develops under normal residential use, [Builder] will repair it. If improper use, such as heavy truck traffic, has caused the condition, repairs will be your responsibility.

Settling. Settling next to your garage floor of up to $1\frac{1}{2}$ inches across the width of the driveway is normal. Settling or depressions elsewhere in the driveway of up to one inch in any eight-foot radius are considered normal. We will repair settling that exceeds these measurements.

Thermal Cracking. Your driveway will exhibit thermal cracking, usually during the first 12 months. These cracks help your driveway adapt to heating and freezing cycles. Cracks should be evaluated in the hottest months—July or August. We will repair cracks that exceed $\frac{1}{2}$ inch in width.

tional details about warranties can be found in BuilderBooks' publications *Warranties and Disclaimers for Builders* and *Contracts and Liability*. You should consult your attorney about your state's requirements. Check whether some of the clauses (especially paragraphs 3, 4, and 8) in the sample warranty that appears in Figure 3.3 would be of value for your company.

FIGURE 3.3 Sample One Year Limited Warranty Agreement

One Year Limited Warranty Agreement

[Builder], hereafter called the "Company," whose office is located at 555 Construction Road, City, State, 55555, extends the following one year limited warranty to:

_____ hereafter referred to as "Owner," who has contracted with the Company for purchase of the home located at

_____ <street address> _____ , Lot _____ , Block_____ , in

_____ County, state of _____ for the purchase price of

$_____ (_____).

The commencement date of the warranty is _____ <Month, Day> _____ , _____ <Year> _____ , and extends for a period of ONE YEAR.

1. COVERAGE ON HOME EXCEPT CONSUMER PRODUCTS

    The Company expressly warrants to the original Owner and to subsequent Owner of the home that the home will be free from defects in materials and workmanship due to noncompliance with the standards set forth in the Limited Warranty Guidelines in effect on the date of this limited warranty, included in the Company Homeowner Manual and which are part of this warranty.

2. COVERAGE ON CONSUMER PRODUCTS

    For purposes of this Limited Warranty Agreement, the term "consumer products" means all appliances, equipment and other items that are consumer products for the purposes of the Magnuson-Moss Warranty Act (15 U.S.C., sections 2301–2312) and which are located in the home on the commencement date of the warranty. The Company expressly warrants that all consumer products will, for a period of one year after the commencement date of this warranty, be free from defects due to noncompliance with generally accepted standards in the state in which the home is located, which assure quality of materials and workmanship. ANY IMPLIED WARRANTIES FOR MERCHANTABILITY, WORKMANSHIP, OR FITNESS FOR INTENDED USE ON ANY SUCH CONSUMER PRODUCTS SHALL TERMINATE ON THE SAME DATE AS THE EXPRESS WARRANTY STATED ABOVE. Some states do not allow limitations on how long an implied warranty lasts, so this limitation may not apply to you. The Company hereby assigns to Owner all rights under manufacturers' warranties covering consumer products. Defects in items covered by manufacturers' warranties are excluded from coverage of this limited warranty, and Owner should follow the procedures in the manufacturers' warranties if defects appear in these items. This warranty gives you specific legal rights, and you may have other rights which vary from state to state.

3. COMPANY'S OBLIGATIONS

    If a covered defect occurs during the one year warranty period, the Company agrees to repair, replace, or pay the Owner the reasonable cost of repairing or replacing the defective item. The Company's total liability under this warranty is limited to the purchase price of the home stated above. The choice among repair, replacement, or payment is the Company's. Any steps taken by the Company to correct defects shall not act to extend the term of this warranty. All repairs by the Company shall be at no charge to the Owner and shall be performed within a reasonable length of time.

4. OWNER'S OBLIGATION

    Owner must provide normal maintenance and proper care of the home according to this warranty, the warranties of manufacturers of consumer products, and generally accepted standards of the state in which the home is located. The Company must be notified in writing, by the Owner, of the

    (*Continued*)

FIGURE 3.3 Sample One Year Limited Warranty Agreement (Continued)

existence of any defect before the Company is responsible for the correction of that defect. Written notice of a defect must be received by the Company prior to the expiration of the warranty on that defect and no action at law or in equity may be brought by the Owner against the Company for failure to remedy or repair any defect about which the Company has not received timely notice in writing. Owner must provide access to the Company during its normal business hours, Monday through Friday, 8:00 a.m. to 5:00 p.m., to inspect the defect reported and, if necessary, to take corrective action.

5. INSURANCE

   In the event the Company repairs or replaces or pays the cost of repairing or replacing any defect covered by this warranty for which the Owner is covered by insurance or a warranty provided by another party, Owner must, upon request of the Company, assign the proceeds of such insurance or other warranty to the Company to the extent of the cost to the Company of such repair or replacement.

6. CONSEQUENTIAL OR INCIDENTAL DAMAGES EXCLUDED

   CONSEQUENTIAL OR INCIDENTAL DAMAGES ARE NOT COVERED BY THIS WARRANTY. Some states do not allow the exclusion or limitation of incidental or consequential damages, so the above limitation or exclusion may not apply to you.

7. OTHER EXCLUSIONS

   THE FOLLOWING ADDITIONAL ITEMS ARE EXCLUDED FROM LIMITED WARRANTY:

   a. Defects in any item that was not part of the original home as constructed by the Company.

   b. Any defect caused by or worsened by negligence, improper maintenance, lack of maintenance, improper action or inaction, or willful or malicious acts by any party other than the Company, its employees, agents, or trade contractors.

   c. Normal wear and tear of the home or consumer products in the home.

   d. Loss or damage caused by acts of God, including but not limited to fire, explosion, smoke, water escape, changes that are not reasonably foreseeable in the level of underground water table, glass breakage, windstorm, hail, lightning, falling trees, aircraft, vehicles, flood, and earthquakes.

   e. Any defect or damage caused by changes in the grading or drainage patterns or by excessive watering of the ground of the Owner's property or adjacent property by any party other than the Company, its employees, agents, or trade contractors.

   f. Any loss or damage that arises while the home is being used primarily for nonresidential purposes.

   g. Any damage to the extent it is caused or made worse by the failure of anyone other than the Company or its employees, agents, or trade contractors to comply with the requirements of this warranty or the requirements of warranties of manufacturers of appliances, equipment, or fixtures.

   h. Any defect or damage that is covered by a manufacturer's warranty that has been assigned to Owner under paragraph 2 of this Limited Warranty.

   I. Failure of Owner to take timely action to minimize loss or damage or failure of Owner to give the Company timely notice of the defect.

   j. Insect or animal damage.

(Continued)

FIGURE 3.3 Sample One Year Limited Warranty Agreement (Continued)

8. ARBITRATION OF DISPUTE

The Owner shall promptly contact the Company's warranty department regarding any disputes involving this Agreement. If discussions between the parties do not resolve such dispute, either party may, upon written notice to the other party, submit such dispute to arbitration with each party hereto selecting one arbitrator, who shall then select the third arbitrator. The arbitrators shall proceed under the construction industry rules of the American Arbitration Association. The award of the majority of the arbitrators shall be final, conclusive and binding upon the parties. The expenses of the arbitrators shall be shared equally, but each party shall bear its own fees and costs.

9. EXCLUSIVE WARRANTY

THE COMPANY AND OWNER AGREE THAT THIS LIMITED WARRANTY ON THE HOME IS IN LIEU OF ALL WARRANTIES OF HABITABILITY OR WORKMANLIKE CONSTRUCTION OR ANY OTHER WARRANTIES, EXPRESS OR IMPLIED, TO WHICH OWNER MIGHT BE ENTITLED, EXCEPT AS TO CONSUMER PRODUCTS. NO EMPLOYEE, TRADE CONTRACTOR, OR AGENT OF THE COMPANY HAS THE AUTHORITY TO CHANGE THE TERMS OF THIS ONE YEAR LIMITED WARRANTY.

DATED the _____<Date>_____ day of _____<Month>_____ , ___<Year>___

_____   _____
Homeowner   Builder

_____
Homeowner

Phone (or Communication) Log

Banish teeny-tiny phone message pads. They are too small to record enough information, are easily lost, and they disappear in the files (if they make it that far). Instead keep a supply of $8\frac{1}{2} \times 11$ phone log sheets like the one shown in Figure 3.4 near every phone. Phone logs offer several advantages because they:

- Provide room to document concerns, action taken, and follow up notes.
- Provide a convenient tool for delegating a situation to another staff member.
- Support good follow up by providing a reminder to check back with the homeowner.

FIGURE 3.4 Phone Log

Date _____

Homeowner _____   Closing Date_____

Address_____

Phone (H) _____   Phone (W)_____

Message   Response

Follow-Up Notes

By _____

- Help keep files complete.
- Can be used in court as business records (be sure to sign and date each one).

Incident Report

Builders frequently describe a series of occurrences with a client that lead to frustration and conflict. The question that follows is usually "What do we do now?" The response almost always begins with "What do you have in the file?" In other words, while what you remember is interesting, the real issue is "What can you prove?" Too many times the answer is "Nothing."

Builders routinely ask employees to "Write that up for the file, will you?" and those employees routinely respond "Sure." Then nothing happens. When company personnel are unsure of how to perform a task, that task is unlikely to be performed with any great speed-if at all. Details are often left to memory for one or more of three reasons: (1) Busy schedules: "I meant to document that. I just never got around to it." (2) Failure to anticipate the potential impact of a conversation or an event: I should have realized that when Mrs. Jones chased me down the street waving a tire iron, that would not be the end of our disagreement. (3) Lack of a convenient system for creating the documentation: I know I should have written that up, I just didn't know where to start.

FIGURE 3.5 Incident Report

Date of Incident _____   Time _____

Location _____

Participants_____   Witnesses _____

Events _____

Signature _____   Date_____

The incident report form in Figure 3.5 offers a solution. An incident report can document anything from a disagreement with a customer to accidentally damaged materials. The description of events is best organized chronologically and can include quotes. Limit remarks to the facts. This is not the place to say "This customer needs more sleep."

Chapter 4

Company Communication

Builders struggle to balance too many meetings and reports against too little communication and information. With the right balance, staff feel that they are a part of things and make better decisions. Unless your company is a one man show, regular communication is essential to keep everyone working for the same goals. This communication usually means meetings and reports, including summaries of responses to customer satisfaction surveys-a report card on how well an organization is meeting its quality and service goals.

Meetings

Business people complain about meetings–often with good reason. Common complaints include that there are too many meetings, they are a waste of time, participants are unprepared or negative, meetings lack clear purpose, do not start (or end) on time, include too many people, not enough people, or the wrong people, or decisions are made and forgotten.

Meetings can be routine, with a set schedule, regular attendees, and standardized agendas or they can be occasional, such as a company task force that addresses a particular issue. Most meetings are for the purpose of sharing or creating information. Before holding any meeting, ask whether an alternate method of communicating would work as well. Figure 4.1 lists alternatives.

When meetings are necessary, make them productive and efficient. Set standards for meetings just as you might for other aspects of your operations. These could include points such as those listed in Figure 4.2. Include those who can contribute, those who would benefit from hearing the discussion, or those affected by the outcome. Where will the meeting be held? Will this location be convenient to the attendees, and appropriate for the size and length of the meeting? Set a date and time and provide sufficient notice so participants can prepare and attend without rescheduling other appointments. Announce the meeting on one sheet including the date, time, location, participants, and questions to be addressed. Hold routine meetings at appropriate intervals-far enough apart that significant new information is covered, but not so far apart that people are making decisions in an information vacuum. Routine meetings should have standardized agendas including room for unusual issues.

FIGURE 4.1 Alternatives to Meetings

- Memo
- E-mail
- Report
- Newsletter
- Bulletin board notice
- Routing sheet
- Phone/conference call
- Public address announcement
- Notice in pay envelopes
- Closed circuit TV
- Taped dictation
- Make the decision yourself

FIGURE 4.2 Quality Standards for Meetings

Meeting Leader
- Remind reception about the meeting and what time you will end or take a break.
- Start on time.
- Ensure the group proceeds through the agenda in an orderly fashion.
- Encourage participation, allow everyone an opportunity to talk.
- Have sufficient copies of materials to be distributed.
- Review notes from the previous meeting and ask for relevant updates.
- As action items are identified, make note of who will follow through when an update is due.
- Distribute the minutes of the meeting within 24 hours.

Participants
- Arrive prepared.
- Turn phones, pagers, and radios off (unless someone is waiting for urgent information).
- Listen carefully.
- Contribute relevant, positive comments.
- Avoid sharing war stories unless they are relevant to the topic under discussion.
- Avoid negativism, whining, and blaming.
- Offer suggestions and solutions.

Common Meetings

Quarterly or annual company meetings are becoming more common because companies are realizing that employees need and want a sense of direction. A typical agenda might include company goals, strategies, progress reports, challenges, new employee introductions and recognition. Other common meetings range from annual management retreats to weekly departmental meetings to daily huddles. For example, the warranty staff should meet regularly, usually weekly in most companies, but at least once a month. Potential agenda topics appear in Figure 4.3. Trade contractor meetings on a regular basis or upon the opening of a new community offer forums to resolve quality or procedural issues-whether trade to builder or trade to trade. In larger companies, cross functional meetings such as community team meetings, involving managers or staff from two or more departments are increasing in popularity.

Community Team Meeting

A relatively new type of meeting, the community team meeting is gaining in popularity as builders see benefits accrue. The random calls or visits that occur daily between sales and construction address questions and concerns raised by home buyers on an equally random basis. While urgent questions should be discussed by the on-site team as they arise, routine (usually weekly) community team meetings provide for comprehensive review of every customer's status and allows orderly operations. Home buyers quickly discover that issues they discuss with one team member are conveyed to the others. This increases the buyer's confidence in the team and discourages that occasional dishonest customer from attempting to play team members off against each other.

The sales person and the superintendent meet and review each buyer's file. This is a time to share new information ("The Thompson's' mortgage has been approved"), provide updates on recent contacts with the buyer ("Mrs. Harrington is worried about the dying tree on her home site. Do we have any plans of removing it?"), or confirm new change orders (Mr. Smith wants to go ahead with the fireplace. He'll be out today to sign the change order and give us a check.") Target delivery dates should be reviewed.

If a customer has had no contact with their sales person or superintendent over the preceding week, an update can be planned. Depending on the circumstances, sales may put in a call or send an email or fax. Sometimes contact from the superintendent might be more appropriate. Although such updates need not take a lot of time, they can forestall many a crisis and are especially appreciated by out of town home buyers.

If issues arise that require input from other company personnel these are likewise divided between the sales person ("I'm meeting with the loan officer to see if we can rescue the Jones deal.") or superintendent ("This change order price seems out of line; I'll check with purchasing about it.") for resolution. By noting

FIGURE 4.3 Potential Agenda Topics: Warranty Staff Meeting

Company News

Communities

- Community meetings
- Sales staff
- Show homes
- Amenities
- Homeowner association/common

  areas
- Grand opening/Close-out
- Construction staff
- Product changes
- Selections/change orders
- Orientations

Warranty

- Emergencies
- Inspections
  - 60 day
  - 11 month
- Miscellaneous
- Out of warranty
- Pending work orders
- Current
- Expired (over 10 days)

Trade Performance

- New trade orientation
- Evaluations
- Back charges

Accounting

- Approved bills
- Expense sheets
- Gas cards
- Van maintenance
- Budget review

Technology

- Phones, pagers, radios
- Computer software

Staff

- On call schedule
- Vacation schedule
- Shirts

Training Segment

- Homeowner manual review
- Manufacturer rep
- Trade workshop

Competitor Visits

on the agenda who is working on which issues, the topic can easily be revisited at the next meeting. This keeps the details under control.

The condition of inventory homes, model maintenance needs, landscaping and common area details, as well as job-site conditions and trade contractor relations might also be part of this routine review. Weekly on-site meetings are an excellent time for construction to advise sales about changes in methods or materials. Figure 4.4 shows a form to document this.

Although warranty staffing levels may not permit weekly attendance, the warranty rep for the community should sit in on this meeting at least once a month. Design, mortgage, or closing personnel might sit in on some of them as well. A predictable and reliable opportunity for all parties to discuss community issues can prevent details from getting lost. A standard agenda brings consistency to this procedure company wide and from week to week within each community. Potential topics appear in Figure 4.5.

Reports

Well-conceived and effectively used reports help increase customer satisfaction and decrease expense. They should be designed to measure precisely what needs to be measured and keep everyone informed. Reports increase in significance as the company adds more personnel and even for small volume companies, objective data has real value. Find a balance between burying folks in reports and having them function without the information needed to make good decisions.

Benefits

Data drives change. Accurate, objective reports help you convince others where changes are necessary. The full impact begins when you have accumulated enough information for trends or patterns to appear. For instance:

- Information on product performance influences future design and purchasing decisions.
- Repair data may call attention to inadequate supervision or unwise selection of trades.
- Summaries of warranty work orders allow staff work loads to be judged more accurately.
- Completion reports quantify problems with trades, making resolution easier.
- If buyer expectations conflict with promised standards, this can show up in reports of items denied.

Format and Timing

Reports should be readable, consistent, and clear. Begin with the report name and date at the top of each report. Decide on an appropriate format for the data. Whether statistical data is reported in tables or editorial comments from home-

FIGURE 4.4 Construction Notice of Change

Date	
To	Sales, Purchasing, Warranty

From	Construction	<Name>	<Extension>

Re	<Plan Name or Number>

PLEASE ACKNOWLEDGE receipt of this notice by return fax. <FAX #>

Please note the change described below and consider the impact it might have on buyers currently under contract. Take appropriate action to prevent any misunderstandings so that our buyers know what to expect and will be satisfied with their new homes. Please contact me immediately if you anticipate any problems or have any questions.

Describe the pending change:

By _____

Received _____ (date)

FIGURE 4.5 Potential Agenda Topics: Community Team Meeting

Home Buyer File Review (in alphabetical order)
- Status of home and delivery date target
- Change orders or selection issues
- Questions or concerns from the home buyers
- Meetings scheduled with the home buyers

Community Issues
- Condition of inventory homes
- Show home and sales office maintenance
- Landscaping
- Amenities status
- Changes in construction (Fig. 4.5)
- Promotional or marketing update
- Homeowner association items
- Warranty work and trade issues

owner surveys are reported in a text format, present the information in a logical order with clear headings and subheadings to make it understandable.

Beware of reports that duplicate effort without producing any new information. Conduct an inventory of reports routinely produced and how they are used. Answering questions such as those listed in Figure 4.6 can help you identify gaps or duplications in your reporting systems. Most reports are compiled on a weekly (work order completions) or monthly (recurring items) basis. Some are more appropriate quarterly (budget). Of equal importance is completing each report on a timely basis so that when staff review the information, reacting to it is relevant.

FIGURE 4.6 Report Analysis

- What activity is the report about?
- What does the report measure?
- At what interval?
- What is the cut off for data included?
- When is the report due?
- What format best suits the information?
- Who will compile the report?
- Who will receive the report?
- When will the report be discussed?
- What results are expected?

Customer Feedback

From home buyers through real estate agents and trade contractors, no one is in a better position to tell you what it is like to do business with your company than the people who have done business with your company. Most companies are shocked by the responses they receive when they begin using surveys. Examples of survey materials can be found in Section 5 of BuilderBooks' publication, *Dear Homeowner.*

Written Surveys

Maximize responses to your written surveys by keeping the survey brief–one to four pages. Include specific questions and space for general comments. Signatures should be optional, but requested. Expect a 30 to 60 percent rate of return. Including a self-addressed, stamped envelope makes returning the survey more convenient and therefore you receive more replies. Mention a requested return date in your cover letter. Seven to ten days is a reasonable time frame. Including or offering an incentive will improve your rate of return. One builder provided a lawn treatment for homeowners who returned surveys. This gave the homeowners a gift, made the community look great, and provided the lawn service with an opportunity to gain new clients. Sending a thank you gift upon receipt of the completed survey requires that your customer sign the survey.

Consider implementing a two-part written survey. Because the process of purchasing a home takes several months and includes so many details, send the first survey soon after closing (within 10 days or so) while these events are fresh in the customer's mind. Conduct the second survey 9 to 14 months after move-in for feedback on floor plan, quality, and warranty service. This gives homeowners time to experience the seasons, holidays, and warranty service. Written surveys can be managed in house or by an outside survey firm.

Phone Surveys

Like written surveys, phone surveys can be done by in-house personnel or by an outside firm. An advantage to phone surveys is that the interviewer can ask follow-up questions to clarify points a customer makes. A disadvantage is that the interviewer's personality can influence the answers. To maximize success with this method, announce the phone survey with a letter a few days prior.

Post Card Surveys

Some builders use post card surveys after each phase of the process: contract, mortgage application, selections, preconstruction conference, and so on. Self-addressed, postage paid post cards can be handed or mailed to customers. If a customer is disgruntled, the sooner you find out the better. The potential for such immediate feedback keeps staff alert to service principles.

Focus Groups

The impact of a focus group is different than that of surveys. The live answers and immediacy of these discussions makes service issues very vivid. While they do not collect the views of as large a number of folks as written or phone surveys can, they are a source of feedback that is inexpensive, quick, and potent. Consider these hints in your planning:

- Keep the group small so everyone has a change to speak. Invite a few more people than you expect to end up with. A good target number is 12 to 16 participants.
- Saturdays are popular for these sessions with Thursday evenings a second choice.
- Two hours seems to be a good time frame. Include a short break.
- Serve light refreshments.
- Appoint a neutral facilitator who can keep the discussion on track.
- Appoint someone else to take notes.
- Have a planned agenda of several key questions. Keep the topics short and clear. You might want to zero in on one aspect of your process, such as selections, or perhaps you will want comments on the entire experience.
- Remember, this is your homeowners' time to tell you what they think. Avoid explanations. Just listen and absorb what they have to say.

Whether you use written or phone surveys, post cards, focus groups, or some other method for gathering customer feedback, the benefits depend on your doing something with the results. Summarize the information you gathered then circulate and discuss it with appropriate staff. This includes the front line personnel and trades. A timely discussion of results generally leads to dramatic improvement.

Interpreting the feedback you obtain from surveys is an art in itself. Discipline yourself to see through emotions to facts and causes. With practice you will be able to spot the difference between a one-time problem–perhaps caused by an isolated clash of personalities or an unusual circumstance–and a fundamental weakness in your system or staff that requires action. Whether you identify one idea or a long list of improvements for your customer service program, the effort will be worth the time.

Chapter 5

Service Policies and Procedures

Builders often react to service issues one at a time, as they arise. The result can be a patch-work-quilt system that produces more "busy-ness" and expense than positive results. Deeply ingrained traditions may be followed blindly long after they have become inappropriate. At the other extreme, shoot from the hip responses are often regretted later and similar situations may be handled differently from one day to the next, depending on who's doing the handling.

A more effective approach begins with recognizing the typical service issues builders face–many common challenges have been identified–and developing guidelines for them before they occur (and while heads are calm). When well-thought out policies and procedures replace the extremes of crisis management or unquestioned habits, an effective service program results. Developing written policies and procedures also clarifies company thinking. As you write down how things get done, previously unnoticed confusion or duplication often becomes visible and can be eliminated.

Documenting Policies and Procedures

To develop effective written policies and procedures, first identify the subject or task you want to define. Suggested subjects appear in Figure 5.1 to get you started (or complement your existing topics). Next decide whether you need a policy or a procedure to address the subject adequately. Policies describe attitude or simple, one step responses. Procedures describe complex behaviors including multi-step processes and paperwork. Samples appear later in this chapter. Develop a policy or procedure when safety, training, consistency, records, or accountability are important. Leave policies or procedures unwritten when they would be impossible to enforce. Before writing a policy or a procedure, ask whether it will help someone do his or her job. Keep policies or procedures as simple as possible. Avoid duplication of effort and unnecessary steps.

Also avoid developing policies or procedures behind closed doors and handling them to your front line staff to follow. Under such circumstances, many employees will drift back to old, familiar procedures as soon as management stops paying attention. Failure to address this real issue is how some companies end up with a "slogan of the day" reputation. Initiatives are announced and presented by management, then promptly forgotten about. To gain cooperation and commitment, involve employees in developing policies and procedures then practice the coaching techniques described in Chapter 1 until the new methods take hold. Participating in discussions that create (or revise) policies or procedures helps staff members understand steps that they might otherwise resist. Especially where procedures cross departmental lines (as from sales to purchasing to construction–for change orders), creating written procedures draws attention to the need for teamwork. Cooperation increases when people understand the big picture and see where they fit in it. As consensus is reached, consistency increases as well.

As your company approaches the task of creating written procedures, consider:

- Who should be responsible for overseeing the project?
- Who should participate in creating the first draft?-How often will the group meet–and where?
- What materials will they need?
- Who will provide administrative support to do computer input and revisions?
- What is the target completion date?
- When the draft is complete, who will review it and within what time frame?
- Is a legal review necessary?
- To whom will the completed materials be circulated?

Organized in a binder, this collection of customer service policies and procedures becomes your company's customer service manual. Add a title page and table of contents and you have a professional looking handbook for consistent service.

Contents

Once you have created the policies and procedures you identified, organize them in a logical order. Chronological arrangement as shown in Figure 5.1 works well. For each phase of the customer relationship, standard methods applied for every customer are followed by policies or procedures that address situations that occur infrequently. Materials shown in Figures 5.2 and 5.9 show useful components. This formatting can be readily followed by staff and is easily revised.

Objectives. To create and maintain a customer focus, begin every policy or procedure by defining its objective(s) in terms of customer behavior.

Procedures. Describe each step in sequence. The italicized word or phrase shows who perform the steps that follow. Note how clearly this approach shows the interdependent nature of service. Gaps, duplications, and contradictions are usually easy to spot and correct at this point. Clear and well-coordinated policies and procedures, applied by caring employees result in efficiency and customer satisfaction.

Materials. As you list the steps, the materials required to perform them become apparent. List these in the order used and add a copy of each following the procedure. Having a procedure without some form of documentation is unlikely.

FIGURE 5.1 Sample Table of Contents: Customer Service Policies and Procedures Manual

(Continued)

FIGURE 5.1 Sample Table of Contents: Customer Service Policies and Procedures Manual (Continued)

Assessment. How will you measure success? Is there a signed document in the file? A schedule complied with or a check to be cashed? Certain behaviors observed or avoided? Are customers responding with desired ratings on surveys?

(Optional) Training Notes. If your company is large or your staff changes assignments frequently, you might also include a list of training activities as another component of each procedure.

Sample Forms. Besides including a copy of each form used for a procedure, consider including examples that are filled out correctly. An arrow, circle, or asterisk on these samples can call attention to critical elements that cause problems on the next person's desk if they are incomplete or incorrect.

Modules. Where steps become extremely lengthy or if the same steps are referenced by several procedures, a modular format is helpful. The procedure itself then simply refers to the module maps out detailed actions or guidelines. The sample in Figure 5.3 shows this technique. Modules are easily removed, updated, and re-inserted without having to reprint the entire section.

FIGURE 5.2 Procedure: Sale of a Show Home

Objectives

Home Buyers

Purchase a show home understand the terms and conditions for the purchase, delivery, and warranty.

Procedures

Sales Counselor

- Explains the pros and cons of purchasing a show home.
- Works with buyers and construction to complete a Sale of a Show Home Addendum, Figure 5.3, listing the items included in the show home purchased and any special conditions affecting those items.
- Obtains approval signature from construction on agreed upon actions.
- Obtains buyers' initials for each listed item to confirm agreement on each detail.

Construction

- Details the home in preparation for delivery according to the terms outlined in the Show Home Purchase Addendum.

Materials

Sale of a Show Home Addendum (Fig. 5.3)

Assessment

1. Show home purchasers are delighted with their purchase and their builder.
2. Disagreements about show home details are avoided between homeowners and warranty.

Formatting

A unified appearance helps those who use the manual see how the parts fit together. If all departments adhere to the same formatting training, cross training, and daily use are easier. Make formatting decisions early to avoid re-working the material. Consider formatting decisions on such issues as:

- Will you need sections, chapters, or both?
- What margin settings are appropriate?
- Will you double space between paragraphs? Avoid "sea of print" pages that contain one long paragraph. These are hard to read and harder still to

FIGURE 5.3 Sale of a Show Home Addendum

The show home located at _____ is offered for sale at a base price of
$_____. This price is based on the home in "as is" condition.

1. The inventory that follows details the condition of the extras (decorator items, landscaping materials, furnishings, and so on) included in the home that were part of [Builder] merchandising plan for the home and will remain with the home upon sale. Purchasers acknowledge they have inspected the home carefully and confirmed the inventory is a true and accurate description of this home except as noted in item 3 below. For each item listed, Purchasers have indicated any action they request [Builder] to take prior to closing and delivery of the home. The corresponding charge, if any, will be added to the price of the home. Items not selected by Purchasers will be delivered in "as is" condition. All maintenance and future alterations to items listed "to remain 'as is'" will be the Purchasers responsibility.

Item	Condition	Accepted "As Is"	Agreed upon Action	Cost	Buyer Confirmation

2. Alterations from the standard floor plan or specifications for this home include the items listed below. Purchasers may have the option as indicated below of having the alteration returned to the standard plan. All decisions listed below are final. All maintenance and future alterations to items listed "to remain 'as is'" will be the Purchasers responsibility.

Item	Builder will Restore per Standard Plan	To Remain "As Is"	Buyer Confirmation

3. Items excluded from this inventory, which will be inspected after removal of furnishings:

4. Builder's standard Limited Warranty is amended as follows:

All terms and conditions are hereby acknowledged and agreed to

_____ _____
Purchaser   Date   Builder   Date

_____ _____
Purchaser   Date   Sales Counselor   Date

FIGURE 5.4 Procedure: 60-Day Builder-Initiated Warranty Visit (Continued)

Objectives

Homeowners

1. Move in confident that attention from [Builder] will continue after they have paid for their home.

2. Perceive that [Builder] stands behind its homes.

3. Benefit from an informed review of their home to confirm that it meets [Builder] standards and to identify warranty items of which they may not have been aware.

4. Have an opportunity to review key maintenance points covered during their orientation and to ask questions about the use and care of their home.

5. Are invited to call attention to potential warranty items they have noticed.

6. Provide feedback on work performed on orientation items to confirm their satisfaction or identify items that may need further attention.

7. Receive feedback and guidance on any errors they have made in landscaping, alteration of drainage that might impact their home, or maintenance oversights in the home's interior.

Procedures

Warranty Rep

1. At the conclusion of the orientation:
   a. Reviews the purposes of the 60-day visit, calling attention to the description of this visit the homeowner manual, page <page #>.
   b. Sets an appointment with the home buyers, noting the day and time on the orientation forms and in his or her appointment book.

Warranty Administrator

2. Upon receipt of copies of the orientation forms, records the 60-day visit appointment for a confirmation call 5 days prior.

3. Mails the confirmation of 60-day appointment letter, Figure 5.5, with another warranty service request form, Figure 5.6, for the homeowner's convenience and a copy of the 60-Day Warranty Visit Checklist, Figure 5.7.

4. Five days prior, contacts the homeowner to confirm the appointment (or re-schedule if necessary), and asks whether the homeowner has noticed any potential warranty items.

5. If the homeowner has noted items, asks that he or she fax or email that list.

6. Notifies the warranty rep of the status of the appointment by completing in the homeowner's name and address on a 60-day warranty checklist, Figure 5.7, and forwards the checklist along with any list of items the homeowner has sent in.

Warranty Rep

7. Conducts the 60-day visit according to the guidelines listed in Figure 5.8, Warranty Inspection Guidelines.

(Continued)

FIGURE 5.4 Procedure: 60-Day Builder-Initiated Warranty Visit (Continued)

Warranty Administrator

8. Within one business day of receipt of the 60-day checklist, issues required work orders.

9. As needed, prepares follow up letter confirming denial of service on maintenance items noted on inspection report or needed corrections to homeowner caused conditions.

Materials

1. 60-Day Warranty Visit confirmation letter (Fig. 5.5)

2. Warranty Service Request (Fig. 5.6)

3. 60-Day Warranty Visit Checklist (Fig. 5.7)

4. Warranty Inspection Guidelines (Fig. 5.8)

Assessment

1. Ninety-five percent of homeowners participate in the 60-day warranty visit.

2. Ninety percent of homeowners rate overall warranty service "excellent" on customer satisfaction survey questions. No homeowner rates warranty service below "good."

<revised date>

---

follow. Spacing breaks information into visual chunks that reinforce the step-by-step nature of procedures.

- Will you use a header or footer?

Decide on font size and style for titles, text, headings, and subheads. (Select no more than two font styles for a clean look. For instance, Arial for titles, Times New Roman for everything else.)

- Do legal size forms need to be reduced to fit conveniently?
- How will pages be numbered? Centered page numbers allow for 2-sided copying. Section numbers (2.1, 2.2, 2.3, and so on) allow for revisions without affecting numbers in other sections.
- Show a "revised date" at the end of each component and on the table of contents to avoid confusion as revisions accumulate.
- You will want a table of contents, but do you need an index as well?
- What cover or binding treatment will provide the desired image and allow for updates?

FIGURE 5.5 Confirmation of 60-Day Warranty Visit

<Date>

<Homeowner Name>
<Address>
<City, State, and Zip Code>

Dear <Homeowner>:

Getting settled in a new home is exciting and a lot of work. We hope your move went well and that you are enjoying your new home. We would like to confirm the warranty appointment we made with you during your orientation for

<date> at <time>.

This meeting has three purposes:

1. We would like to confirm the home we delivered to you is performing to [Builder] standards. A copy of the checklist we review is enclosed.
2. If you have noticed warranty items your believe need attention, please note them on the enclosed warranty service request form. We will review them with you and make a repair determination.
3. We will answer questions you may have about the operation or care of your home.

If you find you need to re-schedule this appointment, please call our office.

We look forward to working with you.

Sincerely yours,

<Builder>

Enclosure

FIGURE 5.6 Warranty Service Request

For your protection and to allow efficient operations, our warranty service system is based on your written report of non-emergency items. Please use this form to notify us of warranty items. Mail to the address shown above. We will contact you to set an inspection appointment. Service appointments are available from 7 a.m. to 4 p.m., Monday through Friday. *Thank you for your cooperation.*

Name	_____	Date	_____
Address	_____	Community	_____
Phone (H)	_____	Lot#	_____
Phone (W)	_____	Plan	_____
Fax	_____	Closing Date	_____

Service Requested   ***Service Action****

                                    Warranty   Courtesy   Maintenance

Comment

*<u>Warranty</u> or <u>Courtesy</u> indicates Builder will correct.
<u>Maintenance</u> indicates a homeowner responsibility.

Homeowner _____

Distribution

Establish a system for distribution of the original materials and future updates (including an updated table of contents with the changed "revised date"). Avoid making the assumption that because an individual does not perform a task he or she does not need information about how it is done. When employees operate in a vacuum, teamwork disappears. Though not everyone needs to be able to perform every job, everyone should be generally aware of how things are done and have access to more details when needed.

Training Tool

Once you have reduced them to writing, policies and procedures provide a tool for training new personnel and cross-training veterans. Routine review of policies and procedures helps everyone understand his or her assignment as well as what others in the company do. Supplement formal "classroom" style training

FIGURE 5.7 Builder-Initiated Warranty Visit Checklist

Warranty Check-up

Name	_____	Date	_____
Address	_____	Community	_____
Phone (H)	_____	Lot#	_____
Phone (W)	_____	Plan	_____
Fax	_____	Closing Date	_____

_____ 60-Day   _____ 11-Month   Inspection Date _____

- Backfill
- Drainage
- Downspout extensions
- Concrete flatwork
- Front door
  - Lock and deadbolt
  - Threshold
  - Weatherstrip
  - Doorbell
- Back door
  - Lock
  - Threshold
  - Weatherstrip
- Patio door lock
- Garage overhead door
- Smoke detectors
- Furnace filter
- Interior doors
- Interior trim
- Cabinets
- Tile
- Caulk
- Window operation
- Drywall
- Floor coverings
- Homeowner list?
  - ___ Yes, attached
  - ___ None

By _____

FIGURE 5.8 Warranty Inspection Guidelines

Preliminaries

1. When applicable, read the homeowner's list prior to the appointment and conduct appropriate research in preparation for the inspection.

2. Review the orientation list, any previous warranty work orders, and consider whether checking with the sales person or superintendent would be helpful.

3. Avoid conclusions prior to seeing the items. Homeowners often describe items inaccurately, so decisions made from behind a desk may be incorrect.

4. Your homeowner manual is an essential tool to carry on warranty inspections. Designate a place in your office for that and other standard items you need for inspections: pen and clipboard (or palm pilot), inspection forms, camera, business cards, tape measure, your community binder containing floor plans, standard feature lists, and trade contractor information.

5. Complete the top portion of the inspection form-double checking the spelling of the homeowner's name, prior to the appointment.

Upon Arrival

6. Arrive on time or up to 5 minutes early. If the homeowner is late, wait a minimum of ten minutes, then leave a door hanger. If you will be late, call as soon as you know, apologize and offer to re-schedule.

7. Park in the street.

8. If you're meeting the homeowner for the first time, introduce yourself, shake hands (unless the customer's cultural background makes doing so inappropriate).

9. Avoid using first names until the homeowner invites you to do so.

10. Chat for a moment if the homeowner's personality makes that appropriate.

11. Introduce the inspection process.

Conducting the Inspection

12. Inspect each item reported to determine its legitimacy for warranty action; check areas where work will be performed for existing cosmetic damage and note such conditions.

13. Add to the list any new items the homeowner points out while viewing the home.

14. Note objectionable conditions caused by homeowner alterations to drainage and explain corrections that are needed to the homeowner.

15. Note for correction any condition that fails to meet [Builder] company standards whether the homeowner points it out or not.

16. Record a decision for each item the homeowner listed: what action is to be taken and by whom? Include details needed to perform the repair efficiently: brand, color, style, size, and so on.

17. Where appropriate, add photos (especially useful with drainage problems), diagrams, and direct quotes from the homeowner to your inspection notes.

18. List follow-up tasks you intend to perform, such as writing a letter advising corrections to landscaping, conducting additional research to resolve an open item, notifying a trade of a back charge, and so on.

(Continued)

FIGURE 5.8 Warranty Inspection Guidelines (Continued)

Conclusion

19. Explain your decisions to the homeowner. In doing this you have two basic choices (match your approach to the situation and the homeowner's personality):

    a. You can explain what action you will provide as you inspect each item.

    b. Look at all the items before committing to specific repairs. Seeing all the items may influence your answers. For instance, after you view the drywall crack in the secondary bedroom, you know that the painter will be in the home and you may as well have him touch up that spot in the entry hall you would otherwise have turned down.

20. Conclude the inspection by explaining what will happen next. Describe the repair process. As necessary, remind the homeowner of the need to remove personal items from a work area.

21. Set a time for an update on unresolved matters, if applicable. Immediately note this commitment on your calendar. Contact the homeowners on time even if it is to tell them you have nothing to tell them yet.

22. Use the homeowner manual to show homeowners the maintenance they need to perform when they have requested "warranty service" on a maintenance item. Rather than say "That's not covered by your warranty," or "We don't fix that" say "This is a maintenance item," or "Let's check the manual and see what we're supposed to do with this. . . . Turns out this is a maintenance item. I'll be glad to answer any questions you have about taking care of it."

23. Sign and date all pages of your inspection.

24. Return the inspection report to the administrator for appropriate follow through.

<revised date>

with reviews at regular staff meetings. On a rotating basis, assign a section for staff to read prior to the meeting so they can note of questions or suggestions to bring up.

Updates and Annual Review

No policy and procedure manual is so thorough that it addresses every potential event. Situations will arise where common sense suggests that policies or procedures be stretched, bent, or ignored completely. Smart companies avoid the pitfall of following rules they created when doing so is unfair to a customer. A unique response, once tested and proven, may set a precedent and ultimately become an addition to the customer service manual.

Ask staff to make notes about changes they want to suggest so inspirations are not forgotten. You might designate a contact person to whom staff can

FIGURE 5.9 Policy: Copies of Warranty File Information

Objective

Homeowner privacy is respected.

Policy

Copies of correspondence and work orders in a warranty file are available to the current owner of a home.

Copies of completed work orders can be provided to a second owner upon receipt of proof of ownership and a letter requesting the copies.

Real estate agents with a listing on a home may obtain copies of completed work orders upon receipt of a letter from the owner of record authorizing us to provide them.

Requests for copies should be filled within 5 business days. Copies can be picked up by the homeowner or mailed.

All conversations over such issues should be courteous, indicating [Builder]'s willingness to cooperate fully in serving the owners of record while protecting their privacy.

<revised date>

forward their suggestions. A minimum of once a year gather use this feedback as part of a comprehensive customer service policy and procedure review–ideally along with a review of your homeowner manual. Consider the following points:

- Does the policy or procedure still accomplish the objectives originally defined?
- Are those objectives still relevant and valuable?
- Has computerization made any steps obsolete?
- Confirm that the steps are still performed by the person or department shown and that the sequence is accurate.
- Double check forms. Have any been revised or new ones created? Can antiquated forms be removed from the manual?
- Are the assessment criteria still accurate and appropriate?
- If your format includes suggested training steps, have any of those changed?

Where you make changes, remember to follow through with a review of those updates with the personnel affected by them. This kind of continuous fine tuning separates high quality organizations from mediocre companies. Over time, this passion to excel leads to the prestigious reputation every good company wants.

Chapter 6

Aligning Customer Expectations

As they shop for the right builder and the right floor plan, buyers are in information gathering mode. They expect to hear new ideas and learn something new. These early stages offer you the best opportunity to influence what customers expect. Every builder has a choice between educating buyers early or arguing with them later.

Answering customer questions accurately is insufficient. Topics the buyers do not ask about need to be mentioned by the company. In fact being the first to mention a subject gives you the advantage of controlling the tone that is set. Particularly where difficult subjects are concerned (change order cut offs, for instance) volunteering your method allows you to avoid defensiveness.

As the process progresses, customer expectations solidify. To complicate this challenge further, buyers cannot tolerate an information vacuum. Where information is missing, customers fill in with assumptions, a past experience, or stories they've heard from others. Adding new information late in the process is unlikely to have much success if incorrect ideas are already in place. Unless they receive good information up front, buyers may perceive completely normal events as problems. Under that circumstance, the most accurate and correct explanations will be seen as excuses. The time to inform customers is before events occur, not during or after. Developing a system that allows this education to occur while the transaction moves forward need not be difficult. Think in terms of:

- **Conversations.** Training and cross training are essential so staff members convey correct and consistent information.
- **Documents and displays.** Double check that show homes, marketing displays, handouts, contract materials, and your homeowner manual also convey correct and consistent information.
- **Meetings.** Approach contract, mortgage, selection, preconstruction, frame stage, orientation, and warranty meetings with unified methods and paperwork.
- **Correspondence.** Use well-timed standard letters to review and reinforce information and assemble a repertoire of letters to address unique situations.
- **Optional.** Some companies have found that home buyer seminars that take advantage of the efficiencies of educating customers in small groups are highly effective.

This dynamic structure for customer communication offers you an opportunity to detail and reinforce essential information. Putting the system to work begins in the sales office at the buyer's first visit.

Sales Initiates Buyer Education

Today's new home salesperson has an educational role that requires technical construction knowledge and an ability to communicate in a forthright manner. By listening carefully as customers describe their housing wants and needs, the salesperson begins to understand the expectations that came through the door with the prospects. Sales professionals watch for and work to correct any expectation a customer has that might lead to disappointment, confrontation, or litigation.

One caution: Companies can grow and improve by meeting the challenge of satisfying honest, rational customers who have high standards. But endlessly striving to satisfy irrational or dishonest customers merely drains company resources. The greatest tools and techniques for aligning buyer expectations will be of little help unless you begin by listening to what prospects want and comparing that to what you offer.

If the prospect in front of you cannot be satisfied with your product, processes, and services, you and the buyer will be better off parting friends now. "Mr. and Mrs. Buyer, after a careful review of your requirements, I find our approaches are different. You will best be served by a builder whose approach more closely matches your own."

Homeowner Manuals

Builders often lament "Buyers don't read the manual!" That's okay–getting buyers to read the manual, while certainly worth their time, is not your goal. Many builders have heard, "I read your manual and I don't care what the schedule says! I expect you to accept this change order." This is similar to drivers who see the speed limit sign and still insist on driving as fast as they want. If buyers read every word of your homeowner manual, that would not guarantee the builder a conflict-free relationship or a satisfied customer.

Rather than trying to get buyers to read the information, your goal is to *establish the authority of the information* in the minds of your customers. To do this, as discussed in Chapter 3, think of your homeowner manual as the conduit into your buyers' expectations. Once your manual is well-organized, written in a friendly tone, and presented in an attractive format, the remaining secret to success is to integrate it throughout the process. When staff members repeatedly and actively use the manual in every interaction with customers, demonstrating over and over that the company follows the manual, customers begin to accept that what the manual says is how company will proceed. The procedures shown in Figure 6.1 provide some suggestions for presentation and staff use of the manual; others are mentioned in later chapters.

FIGURE 6.1 Procedure: Homeowner Manual (HOM) Presentation and Use

Objectives

Home Buyers

1. Have realistic expectations that [Builder] can meet or exceed.
2. Are able to locate answers to questions in the manual.
3. Recognize the value of the manual as a guide now and reference after move-in.
4. Accept the authority of the procedures and guidelines described in the HOM.

Procedures

Sales Counselor

1. Be familiar with all sections of the manual.
2. Display the HOM in the sales center and each inventory home.
3. Maintain a supply of HOMs to give to home buyers.
4. Keep the display copy and all inventory copies updated.
5. Mention the HOM in the sales presentation, for instance, "When you buy one of our homes, you will get a copy of our homeowner manual. It contains a great deal of useful information."
6. Open the contract meeting by presenting the HOM.
   a. Draw attention to the "Purchasing Your Home" section of the HOM.
   b. Present and sign the contract documents.
   c. Briefly review the bulleted summaries at the front of each section of the HOM (3–5 minutes). "I'd like to give you a quick overview so you'll know what to expect."
   d. Suggest that the buyers:
      (1) Store paperwork in the appropriate section of the HOM.
      (2) Bring their HOM to all meetings.
      (3) Store color samples in the HOM for easy reference.
      (4) Read through the next two sections on the mortgage and selection process.
      (5) Note the survey at the back of the HOM. "We're always looking for ways to make this information more helpful."
7. Note which edition of the HOM on the contract meeting agenda (Figure 7.1).
8. Ask buyers to sign the agenda acknowledging receipt of the HOM.
9. Reinforce the importance of the HOM in subsequent conversations with the customer—"Have you had an opportunity to read the homeowner manual? Do you have any questions?"
10. Whenever possible, answer customer's questions about the home buying process by picking up the HOM and showing them the information they need in it. "I believe that's covered in the manual. Yes, here it is on page 29."

(Continued)

FIGURE 6.1 Procedure: Homeowner Manual (HOM) Presentation and Use (Continued)

All Staff

1. Are familiar with the content of the manual and the manner in which it is presented to home buyers by sales.

2. Refer to and follow the manual when working with buyers or homeowners directly, keeping in mind that all decisions are subject to common sense.

3. Bring their manuals to all meetings with home buyers, referring to the material for details or answers to buyer questions. (See individual meeting procedures for more details.)

4. Suggest appropriate additions or changes to the manual for annual revisions.

Materials

1. Display copy of HOM

2. Supply of HOMs for buyers

3. Contract meeting agenda (Figure 7.1)

Assessment

1. Every buyer file contains a signed contract meeting agenda (Figure 7.1) acknowledging receipt of the HOM.

2. The home buyers comply with [Builder] procedures as described in the HOM.

3. The home buyers and [Builder] staff bring their HOMs to all meetings.

4. Questions and conflicts occur infrequently and are readily resolved when they do arise.

<revised date>

Difficult Subjects

If a particularly difficult subject repeatedly causes confusion or conflict, review your existing materials. Is the content adequate? Does it cover the main points in a clear and logical order? Would illustrations (such as digital photos) help establish your points? Once the content says all that you need it to, check on how effectively you are using the information. Look for the teachable moment: When, who, and how can this information best be presented to buyers? Is repetition by another staff member appropriate?

Identify the best place(s) in your process for each subject. Consider credibility and retention and then make certain the person responsible for presenting the information can speak comfortably about the subject. These efforts pay big dividends in customer satisfaction and referrals, prevention of problems, and time saved resolving conflicts.

To emphasize information about subjects that repeatedly cause conflict (care of newly installed lawns, for instance), reproduce the homeowner manual entry that addresses the subject on a no carbon required (NCR) form for buyers to read and sign at contract. Include "from page <page number> of your homeowner manual" on the form. You might also add the topic to one or more of your standard meeting agendas. This technique makes better use of information you already have and eliminates the temptation to develop another version. (See Chapters 3 and 5.)

Client Meetings

Given the amount and complexity of the information involved combined with customers' tendency to miss some details, repetition is essential. The sales person usually has time to cover only the major highlights. Other staff members should add details and reinforcement. This is best done with a series meetings based on planned agendas. With planned agendas, appropriate training can be designed and consistency increases.

Begin with practical decisions that will form the frame work for the meetings, guide your training efforts, and determine what documentation will be needed. Decisions include:

- **Location.** Where will each meeting be held?
- **Attendees.** Who will participate?
- **Agenda.** What topics will be covered?

Ask the personnel who will work with the buyers next what they want the customers to know. These points help form the conclusion of each meeting. "Your next regular meeting will be. . . ." and review the items that will help prepare the buyer for that meeting. Written agendas should list topics t be discussed, items to be delivered, and include room for notes and signatures. Examples appear in chapters 7, 8, and 10. Formatting-logo, margins, font, signatures—should be uniform.

Schedule

Each community may have different scheduling parameters for its meetings. Consider these questions for each type of meeting:

- What event triggers the meeting? The usual answer is one of two things: paperwork has been completed or a particular stage of work on the home.
- Who contacts the home buyers? A checklist of reminders to help the home buyers prepare for each meeting is useful so key points are not overlooked.

- What days and times are appointments available?
- How much notice should you strive to provide to the clients for the appointment?
- Approximately how long will the meeting last?
- Provide the home buyers with copies of the printed agendas; you can include them in your homeowner manual.

Prepare

Staff members who conduct client meetings should prepare in two steps: review of the client's file and talk with other company personnel who've worked with the clients. When possible, avoid making the clients re-tell their story. Seamless service means the company transfers information through the process on behalf of clients. Knowing about questions ahead of time allows personnel to provide the buyer with a status report at the meeting. Clients receive better care and that rare home buyer who might attempt to play one staff member against another has less chance of success. This behind the scenes networking makes the company look attentive and in control.

Conduct

Review these hints with your staff to add polish and prevent problems:

- Arrive early for client meetings; welcoming the clients when they arrive establishes control.
- Begin each meeting by thanking the clients for taking time to meet.
- Introduce each meeting by reviewing its purposes. Show the client the homeowner manual entry describing the meeting. "We've progressed this far in the process." This reinforces that the company follows the manual.
- Cover each agenda item thoroughly, answer client questions, and make note of needed follow up.
- As the meeting comes to a close, check the agenda, confirming that every topic was discussed.
- Conclude each meeting by aligning the home buyers' expectations for the next regular meeting and explaining whom they should contact with questions in the mean time.

Follow Through

Immediately after the meeting-as the buyers are walking out-the person in charge of the meeting should review the notes and decide what steps to take to resolve each. Make tenacious and documented follow through one of the hallmarks of your company. Documenting each resolution is critical to preventing misunderstandings. Should a conflict occur, documentation may resolve it and, ultimately, can be critical to your company's defense.

Inventory Sales or Out-of-Town Buyers

When you sell an inventory home or your customers live out of town, those home buyers have less input and less opportunity to get to know your company. All buyers should receive all literature-even if the events described are already past (such as the opportunity for a frame stage tour when the home sold is an inventory home). Abbreviated versions of normal meetings can go a long way toward creating confidence and trust (tour another home at frame stage to point out the quality your company puts inside the walls). When buyers live out of state and are unavailable for any or all of the normal meetings, make an effort to help them feel involved with updates by fax, phone, or even emailing digital photos. Build the relationship at the same time you build the home.

More details and hints for client meetings appears in the following chapters and you can find still more (as well as a useful training tool) in BuilderBooks' publication, *Meetings with Clients*.

Letters Reinforce Expectations

In addition to your homeowner manual and scheduled meetings, well-timed letters that highlight key points offer another means of communicating with buyers throughout the process. Routine letters can update buyers on construction progress, remind them of upcoming events, or confirm appointments. For example, after signing a purchase agreement, send a thank-you-for-selecting [Builder] letter like the one in Figure 6.2. This sets a professional tone and offers an opportunity to reiterate the next steps in the process. This letter repeat information covered in the homeowner manual and discussed at the contract meeting.

Another example of this technique can be seen in Figure 6.3 in the letter confirming the appointment for the preconstruction meeting. The letter repeats points covered in the homeowner manual and in conversation when the appointment was set. (A sample procedure and agenda appear in Figures 7.4 and 7.5.) More letters appear in the following chapters and an extensive selection can be found *Dear Homeowner,* published by BuilderBooks. These letters do not take the place of personal communication; they reinforce it, helping your buyers stay comfortable by knowing what's coming next.

Home Buyer Seminars

Builder personnel repeat the same information over and over. The more homes you sell and the more attention you give buyers, the more hours are invested in this valuable activity. At some point the time demands become overwhelming and more staff must be added to keep up, unless another approach can be found. That other approach may be home buyer seminars.

Consider organizing one or more classes where buyers hear in small groups the generic information all buyers need to know. Now instead of staff spending

FIGURE 6.2 Letter: Thank You for Selecting [Builder]

<Date>

<Home Buyer Name>
<Address>
<City, State, and Zip Code>

Dear <Home Buyer>:

Congratulations on your decision to build a new home. [Builder] is honored that you have selected us as your home builder.

As your sales associate, <sales associate first and last name>, has explained to you the next steps in this exciting process include finalizing arrangements for financing and making your selections. <Sales associate> can assist you in coordinating these appointments. You can also review detailed information about the mortgage application and selection processes in your Homeowner Manual.

While you are finalizing financing and making selections, we will begin the permitting process, schedule trade contractors, and order materials. Three conditions need to be met in order for us to begin construction. Obtaining a permit is our responsibility; mortgage approval and selections are your responsibility. We will assist you in any way we can with both of these tasks.

Weather permitting, we are normally ready to begin construction of a new home four to six weeks after signing the purchase agreement. We work with a target delivery date during the early months and will update you about progress. You are also welcome to check the status of your home by contacting <sales associate>.

The purchase of a new home is a decision that represents an investment of your emotions, money, and time. We will ask you to participate in several meetings—mortgage, selections, preconstruction, frame stage, orientation, closing, and warranty. We respect that these appointments require you to interrupt your normal schedule and will do everything we can to conduct them efficiently. Each meeting contributes something of value.

As questions come up—and they are inevitable—please contact <sales associate> for assistance. We use a written system through our sales offices to track and document home buyer questions. This assures you a timely response and supports our follow through on commitments we make.

We look forward to working with you through the building process and to having your family as a member of the <community> community.

Best wishes,

<Builder>

FIGURE 6.3 Letter: Preconstruction Conference Confirmation

\<Date\>

\<Home Buyer Name\>
\<Address\>
\<City, State, and Zip Code\>

Dear \<Home Buyer\>:

Your new home's building permit, financing, and selections are now in place. Construction is about to begin and we know you may have many questions. To provide home buyers an opportunity to meet the superintendent who will be in charge of building their new home and hear the answers to their questions, we offer a Preconstruction Conference.

This confirms our appointment with you for your Preconstruction Conference, on

\<date\> at \<time\>.
We will meet at the sales office.

\<Sales associate\> and \<superintendent\> will both attend. Your real estate agent, \<agent name\>, is welcome to observe although we do not require \<his/her\> attendance.

The purposes of this meeting are to review the selections you made, answer questions you have, and provide an overview of the construction process. Unless an error is discovered between the superintendent's specifications for the home and your selection forms, further changes are inappropriate.

A copy of the agenda is enclosed for your review. Please allow one hour for this meeting and keep in mind that a site visit is part of our normal procedure—many buyers bring their camera.

Prior to this meeting, please list questions you have so that we can address them. You will find information about the construction process in "Construction of Your Home," Section 4 of your [Builder] Homeowner Manual.

If you have any questions, please contact me.

Sincerely,

\<Builder\>

Enclosure

two hours with each of 12 home buyers, they spend two hours with a group of 12 and have many hours left over to address unique issues. Individual conversations can follow the general meeting or be scheduled for a later date. Another benefit is that when home buyers hear information (your concrete is likely to have come cracks) with a group of people, they often are more accepting of it. Seeing several others buyers in the room knowingly nod their heads confirms that the information is true. Buyers enjoy getting to meet future neighbors and have an opportunity to get to know your staff in advance of the time that they will actually work together.

The content of the seminar can be based on the information in your home-owner manual, might include a handout and slide presentation, and involve several staff members who each contribute 10 to 15 minutes of instruction. Where difficult subjects are concerned, wording can be rehearsed to achieve a forthright but still friendly presentation of the information. Other company personnel can sit in occasionally and learn from hearing this how to respond correctly to buyer questions. Once the meeting agenda is polished, consistency is assured.

Builders using this method report a noticeable difference between buyers who attend and those unavailable. Home buyer classes may be a part of your future efforts to educate clients.

Chapter 7

Preliminaries

Hemming, hawing, and the universally troublesome "Maybe, I'll have to check" encourage buyers to push boundaries. Prevent this by establishing clear policies and procedures then training your staff to present them in a positive tone. While this approach needs to begin with the first conversation, tangible evidence of the effort comes at the signing of the contract.

Contract

While many of the details covered in purchase documents seem dry and uninteresting when buyers are thinking about carpet and tile selections, making the terms of the agreement clear is important to a healthy relationship. Sales people should candidly address subjects known to cause confusion or controversy, such as what needs to happen before construction begins. Such items can be listed on the agenda for this meeting. Consider the sample in Figure 7.1.

Sales people should also be ready with definitive answers for less common but predictable questions. For example, supposing the buyer of a production home requests a complete set of blueprints. Saying "They don't usually give those out but I'll ask for you" leaves the buyer expecting to get the plans. Later the sales person may be reluctant to deliver a no answer, leading to either resentment over the "company's silly policies" by the sales person or a delay in getting the answer to the buyer–or both. A better response is "We view those as intellectual property. Considerable time and money went into their development-like the secret formula for Coca-Cola. You're welcome to look at them but we retain all copies." This is a matter first of setting a policy about the distribution of your house plans and second of training staff accordingly.

Mortgage

Whether the lender is part of the builder's organization or an independent firm, buyers expect attentive service. Common complaints from home buyers about this aspect of the home buying process include lack of communication during loan application processing and misunderstandings about fees and loan locks. Builders can ensure customers receive excellent treatment by identifying area lenders whose procedures avoid these service failures. Look for mortgage companies that thoroughly review financing options, schedule updates so buyers know what is happening each step along the way, and use a documented system

FIGURE 7.1 Contract Meeting Agenda

<Builder Logo>

Contract

Name	_____	Date	_____
Address	_____	Community	_____
Phone (H)	_____	Lot#	_____
Phone (W)	_____	Plan	_____
Fax	_____	Target Closing	_____

Please confirm you have received or discussed all items checked below.

Documents/Discussion

- Homeowner manual, _____ edition
- Purchase agreement
- 8½ x 11 plans
- Specifications
- Options list for floor plan
- Options displayed in show home list
- Take a Closer Look!
- Homeowner association documents/design review
- Deposit receipt
- Draw/payment schedule
- Mortgage application
- Selection hints
- Construction meetings
- Site visit policies
- Change orders
- Delivery date target
- Orientation
- Closing
- Maintenance hints
- Limited warranty specimen

Follow Up Items

_____
_____
_____
_____
_____
_____
_____
_____
_____
_____
_____
_____
_____
_____
_____
_____
_____
_____
_____
_____
_____
_____
_____
_____

Purchaser	_____	Date _____
Purchaser	_____	Date _____
[Builder]	_____	Date _____

for confirming rates and locks. If your company has a mortgage division, develop an agenda for the loan application meeting.

Selections

In recent years builders at every price point have expanded the choices available to buyers, but not all companies have expanded their efforts to guide the customer through the decision making. While buyers love having choices, they can be overwhelmed by the task. Prepare your home buyers for the numerous decisions they will make and get them thinking early by including practical hints about the process in your homeowner manual. Then make certain that your selection deadlines and policies are reasonable.

Timing

Most purchase agreements obligate home buyers to complete the selection process within a specified time frame. Builder frustration mounts when some home buyers fail to comply with that time frame, especially when the same home buyers are who are late with selections complain about "late" delivery of the home. If only an occasional buyer–such as one out of 30–is late your time frame is probably acceptable and you need to manage that rare customer. However, if 28 out of 30 buyers miss your deadline, some adjustments are probably appropriate. Consider these questions:

- Is the time you allow reasonable for the number of choices you offer?
- Does your contract define the selection timetable and repercussions of late selections?
- Was information buyers needed (especially pricing) to finalize choices readily available?
- Were the buyers adequately prepared for the number of decisions needed to make?
- Do you provide browsing time so buyers can consider combinations without taking staff time?
- Are appointments with selection personnel long enough, but not too long? Buyers often re-think decisions made in haste or when tired, then ask for changes later.
- Can you break the selection schedule into segments, each with its own deadline. For example, identify selections that are required prior to the preconstruction conference and separate them from those that could be finalized by the start of framing.

Standard and Optional Items

Avoid any hint of vagueness here. Document clearly and in a positive tone which optional items you have displayed in each show home. "The following items are displayed in this show home and are available in your home. See your sales person . . ." Review this list at contract and obtain a signature confirming you have done so. This is important enough to list on the selections meeting agenda, as shown in the sample in Figure 7.2.

Consider providing buyers with a list of available options for their plan and ask that they initial "ordered" or "declined" for each item or category. Absolute fury erupts when homeowners believe other buyers got something they did not or that the company promised something in the show home and failed to deliver the item. Update this menu of popular options regularly based on feedback from sales or design personnel about what customers are looking for as well as fluctuations in costs.

Custom Changes

If your company accepts custom change requests strive to return the pricing within 48 to 72 hours. Prevent conversations along the lines of "If I'd known you could do that I would have ordered it, too," by alerting home buyers: "Please keep in mind that your new neighbors have this same opportunity and may request still other features. We make no claim that we mention or offer every possible idea."

You can minimize "kid in a candy store" syndrome, which results in buyers submitting dozens of custom pricing requests and then buying few if any by advising them of this policy:

All requests for custom features require a custom design/pricing deposit of $200. The full amount becomes a credit against the cost of the change if you approve the change order. If you decide not to proceed, [Builder] retains the design/pricing deposit.

To be fair, you might respond to the first three such requests at no charge, then apply the design deposit. (Change orders are also discussed in Chapter 8.)

Quality

Differences in standards between the home buyers and their builder are common. This seldom causes a problem when the builder's standards are higher than the home buyers' but does become a concern when the situation is reversed. Regardless of how exceptional your quality is, some buyers will have standards higher than those you meet in your homes. This is where the benefit of aligning expectations early shows dramatically. If the buyers knew before they committed to purchase the home that something varied from their personal preference, the issue would have already been resolved. Figure 7.3, Take a Closer Look! helps get customers in touch with the quality you are offering and they are buying.

Sweat Equity

Buyer visits to your selection center can result in sticker shock. When buyers ask whether they can supply materials ("We can get that at <store> for half the

FIGURE 7.2 Selection Meeting Agenda

<Builder Logo>
New Home Selections

Name _____	Date _____
Address _____	Community _____
Phone (H) _____	Lot# _____
Phone (W) _____	Plan _____
Fax _____	Target Closing _____

Please confirm you have received or discussed all items checked below.

Documents/Discussion   **Follow Up Items**

- Standard features
- Optional featuers
- Custom features
  - pricing
  - cut off dates
  - design deposit
  - administrative fee
- Selection hints
  - browsing hours
  - informed choices
  - allowances
  - color variations
  - exterior choices
  - selection hold
  - availability
  - record of selections
  - worksheet
  - final appointment
- Selection forms
- Next: Preconstruction
  meeting

Purchaser _____   Date _____

Purchaser _____   Date _____

[Builder] _____   Date _____

FIGURE 7.3 Check Out Our Quality

Take a Closer Look!

[Builder] is proud of the quality we deliver in our homes. This quality is clearly displayed in our show homes. The standards demonstrated satisfy our critical eye, but it is your satisfaction that matters. Therefore we invite you to confirm that those standards satisfy your critical eye. Because we intend to build your home to the same standards displayed in our show homes, we invite you to visit at length and examine the details for yourself. Your satisfaction with the quality of the home you purchase is so important to us, we will ask you to acknowledge that you have fully examined our work by signing this sheet prior to our selling you a home. Take your time and take a closer look!

- Floors and walls
- Drywall
- Insulation
- Interior trim
- Stairs and rails
- Paint and stain
- Windows
- Exterior doors
- Floor coverings
- Cabinets
- Countertops
- Plumbing fixtures

- Tile or stone
- Electrical fixtures
- Heat and air system
- Fireplace
- Appliances
- Hardware and fixtures
- Water heater
- Garage
- Exterior finish
- Decks
- Concrete flatwork
- Roof

To avoid surprises later, we suggest that you sit on the floor and look under countertops, window sills. Step into closets and see the back of the doors and interior finish. Walk to the corners of rooms, turn and look. Are our walls plumb enough to please you? Floors level enough? Listen to the home as you move from room to room. Open and close doors and drawers. Check our fit and finish as well as the materials included. If you have any questions about what you see—mechanical, philosophical, technical, or trivial—we will be glad to get you an answer.

Although perfection eludes us, we are committed to a standard of excellence that pleases our buyers. If you agree that we have achieved that level of quality, we'd like to sell you a home!

Acknowledgment
I/we have taken a closer look and approve of the quality and value offered by [Builder].

Purchaser	Date	Purchaser	Date

cost.") or perform labor ("Filling nail holes and sanding interior trim is one of my hobbies."), builders often cringe. While these sweat equity arrangements appeal to some buyers, the jury is still out, however, as to who does the most sweating, the buyers or the builder. Builders willing to enter sweat equity agreements have learned to clarify responsibilities in a written addendum that defines responsibilities for such items as materials and labor, scheduling, loss or damage, work site conditions (trash removal and so on), meeting applicable codes, insurance, and warranty coverage. Get assistance from your attorney to create an appropriate document. Or better still, if your choice is to avoid sweat equity arrangements, document your position in your homeowner manual:

[Builder] is a single source company. That means that we select all personnel and companies who will contribute to your home. We order all materials and products from suppliers with whom we have established relationships. Although sweat equity arrangements are unavailable as a part of our purchase agreement, you are welcome to add your personal touches to the home after you close and take possession of it.

Then make certain that all staff can comfortably say "We are a single source builder. That means . . ." (Again, the word no was avoided.)

Preconstruction Meeting

Normal events in the construction process can alarm uninformed home buyers and erode their confidence in your company. Good communication prevents much of this. Just as show homes and homeowner manuals help align home buyer expectations about your product, the preconstruction meeting aligns expectations about the *building process.*

Trust and confidence in the person in charge of building the new home allows home buyers to relax. In a practical sense, the preconstruction conference offers everyone an opportunity to double check the accuracy of paperwork. Misunderstandings or omissions are easier to correct while they are still on paper than after they appear in the home. If any questions exist in the clients' minds, now is the time to address them.

Preconstruction conferences are usually held after most decisions about the home are made and just before construction begins. The home buyers, sales person, and the superintendent should attend. In most companies the sales person sets up this meeting. The letter in Figure 6.3 confirms the appointment. Figure 7.4 outlines the procedure and a sample appears in Figure 7.5.

FIGURE 7.4 Procedure: Preconstruction Meeting

Objectives

Home Buyers have a clear overview of the construction process and a high level of confidence in [Builder].

Procedures

Sales Counselor

1. Schedule the preconstruction meeting with the superintendent and the home buyers.
2. Mail confirmation letter (Fig. 6.3).
3. Prepare for the meeting by reviewing the file with the superintendent.
4. Note items that need follow up attention.
5. Follow up on any contract related issues that remain after the meeting.
6. Offer suggestions for revising the standard agenda.

Superintendent

1. Prepare for the meeting by reviewing the file with the sales counselor.
2. Guide the meeting according to the agenda.
3. Follow up on any construction issues that remain at the end of the meeting.
4. Offer suggestions for revising the standard agenda.

Materials

1. Preconstruction conference confirmation letter (Fig. 6.3).
2. Buyers' contract file.
3. Preconstruction meeting agenda (Fig. 7.5).

Assessment

1. Sales agent, buyers, and superintendent agree regarding features and details selected for the new home.
2. Any questions that arise during construction are answered quickly and easily using well-established lines of communication.

FIGURE 7.5 Preconstruction Meeting Agenda

<Builder Logo>

Preconstruction Meeting

Name _____	Date _____
Address _____	Community _____
Phone (H) _____	Lot# _____
Phone (W) _____	Plan _____
Fax _____	Target Closing _____

Please confirm you have received or discussed all items checked below.

Documents/Discussion   **Follow Up Items**

- Site plan
- Soil report
- Drainage plan
- Status of permit
- Utilities status
- Landscape plans
- House plans
- Specifications
- Selections and options
- Change orders
- Change order cut-off
- Target start date
- Schedule/sequence
- Events that extend schedule
- Target delivery date
- "Nothing's happening"
- Quality, inspection of work
- "Something's missing"
- Questions
- Next step: Frame stage meeting

Discussion: On Site

- Lot boundaries, easements
- Orientation of home
- Trees, natural features
- Drainage
- Mailbox utility junctions

Purchaser _____   Date _____

Purchaser _____   Date _____

[Builder] _____   Date _____

Chapter 8

Customers and Construction

Builders sell a process as well as a home. Homeowner surveys often show considerable frustration with that process. Criticisms include lack of communication, lack of responsiveness. Friction over common aspects of the building process such as change orders, site visits, or delivery dates may arise with some buyers as well. These complaints and others like them can be avoided with carefully planned procedures, thorough training, and good customer communication techniques.

Change Orders

Everyone has the perfect house in his or her mind. Almost any customer can look at a floor plan and think of things to change. In today's new home market, the willingness to let the customer initiate changes has practically become a required feature, regardless of the price range. The boundaries between "production" and "custom" have definitely become blurred.

Agreeing to make changes and then making them badly is worse than saying no to them in the first place. The marketing advantage that comes from accepting changes can easily become a detriment to future sales if the construction system handles these changes poorly. Successfully managing this while achieving the promised quality, on time and within budget, takes careful planning and effective communication among staff, buyers, and trades.

Types of Changes

Recognize that change orders fall into three broad categories. Some change a structural detail of the floor plan, customizing it for the buyers' needs. Others add or delete a standard of optional item outside the normal selection schedule. And a third type involves the buyers changing a selection they made earlier, as from beige to white carpet. This recognition is worthwhile in that you may want to establish policies according to the type as well as the timing of the change request. For example, perhaps structural changes carry the earliest deadline. Features may come under a slightly later deadline, and changes to finish material or colors might be possible until the material is actually ordered.

Timing of Changes

Cut-off dates are the common control builders (attempt to) use to manage change orders. Set these limitations in a friendly manner. "Changes in <components> are welcome until <stage of construction>. In determining your particular schedule, set the cut-off points to allow for some delay from customers. This cushion allows you to be friendly when they make final decisions late.

Back this up with effective training to stop late requests where they start. Salespeople and design center staff need a conversational knowledge of the consequences of change orders so they can explain them to buyers: A momentum is created when buyers make the decision to build a home. A start order goes to the field and orders are sent to trades and suppliers. Company personnel, trades, and suppliers respond by ordering the materials they will need and scheduling the work to be done.

As the process continues, more people (and schedules) are involved. Some components have long lead times and others are special ordered. Some are shipped long distances. One or more tradesman may need to be scheduled far in advance of the actual installation. Changing the direction of this process once it is moving is difficult. The further into the schedule that changes are accepted, the more likely those changes can affect costs, schedules, quality, or all three.

Processing

A processing concern is lack of sufficient detail on the initial change request. Salespeople or design personnel need should include diagrams, drawings, catalog pages, model number, location information, size, color, and so on to define exactly what the buyer wants. Avoid the frustration experienced by the superintendent who received a change order instructing him to add a window to the living room with no information about size, location, or type of window. Completing the form with detailed and accurate information helps everyone.

Administrative Fees

Many builders charge administrative fees for change orders requested after they begin construction, or after the home has passed a designated point in the building process. Some simply use 10 to 30 days after signing the contract, others use a specific event, such as the start of framing. Avoiding this extra charge can be an incentive for some buyers to make changes early.

Some companies offer a "no change discount" ($1000 is typical), which is credited at closing for buyers who make no changes after a set point in the process. Other buyers, especially those building high-end homes, are undaunted by fees of even $200 to $500 per change order. Builders do considerable legitimate work and take on significant risk when they accept late changes. Compensation should be reasonable and consistently collected. A builder philosophy that says "If the buyer has the time and the money, we'll change whatever they want us to" is fine if future buyers do not end up paying for expenses created by previous buyers stretching the system.

Quality

Change requests involve features and details that are important to customers. Time is invested thinking (dreaming) about them, having meetings to fine-tune them, and filling out change order forms. Customers usually pay extra for them. After all that, it is to be expected that buyers pay close attention to changes. Customers can be highly critical if the end results are disappointing.

When you commit to changes, you must also commit to delivering those changes at the same quality level as the rest of the home.

Using a trade, a product, or a method with which you have no experience carries some risk. Which trade can be depended on to do high quality work? What other components of the home will be affected by the new item? What preparations are necessary for using glass block? What are the implications for painted window grids if a hot tub is installed in a sunroom? What waterproofing steps are needed if a deck is built over a garage? What are the warranty implications of such changes?

Design details that have not been field-tested are more likely to produce surprises. When one home buyer doubled the size of the standard patio, no one anticipated the need to alter normal gutter installation. The downspout emptied in the middle of the patio. Conflict followed when the builder wanted to collect for corrections. The home buyer felt it was the builder's job to anticipate such details. After all, the builder is the professional in this relationship and can fairly be assumed to have superior knowledge by the home buyer.

Successfully working with unfamiliar items often requires research beyond determining the hard costs. Knowing how to fit the change comfortably into the original plan, what warranty services might be needed, and what price to charge to make a reasonable profit are important details.

Frame Stage Meeting

Frame stage tours generally involve the home buyers and the superintendent, take place in a home under construction where noise and dust are likely but sitting down is not, and usually take from 20 to 40 minutes. Because of these informal circumstances, the value of the frame stage meeting can easily be overlooked. Avoid this by considering what this meeting means to the home buyers.

This is the home buyers' opportunity to have your undivided attention, to ask questions about their home, and have you point out the quality features your company builds into the walls of their home. In training your personnel for the frame stage meeting, remind them to prepare for by studying the buyer's file and check with the sales person for any updates about the home buyers. An additional and vital step is getting the home site ready. Steps might include checking on the following:

- Delivered materials are stored neatly and securely. These are the components that will eventually make up the home; the client will want to see them being protected from damage.
- Scraps and trash have been removed from the site and the home itself swept.

- How does the surrounding area look? Muddy streets, tilting portable toilets, and weathered stacks of unused materials should be corrected.
- Safety considerations have been addressed, appropriate rails installed and other protective measures taken.
- Compare the home to the information in the file. Is anything missing or incorrect?
- What is the status of any issue the home buyers have raised?
- Do you have any questions for the home buyers? If so, note them on the meeting agenda.
- Radios that can be heard in the next county, squealing tires, and screamed obscenities set an "anything goes" tone on the job. Buyers are reassured when details show someone who respects the work in control. Everything counts.

Finally, discuss the common tendency is to show up for the frame stage meeting and stroll around the home more or less haphazardly. Think of the advantage your staff will have at the orientation if your home buyers are accustomed to a standardized itinerary. Formalize the frame stage tour by following a route that closely matches that of the orientation yet to come. Greet your clients at the street when they arrive, tour the exterior of the home first, enter through the front door and end in the future kitchen. Reasons for this itinerary are described in Chapter 9.

Upon entering each area or room, the superintendent should call attention to optional or custom features the home buyers ordered. Next, confirm that the number and locations of phone, cable, and electrical outlets are correct. Last then offer general information about standard quality techniques the company uses in every one which are visible in the room. An example of an agenda for this meeting appears in Figure 8.1.

Random Site Visits

In addition to the frame stage meeting, most home buyers visit the job site regularly, in some cases, continuously. They do this for one or both of two reasons: excitement about the new home ("Is it done yet?") or lack of confidence in the builder ("Did you see *Dateline*? Boy, you've got to watch them every minute!"). The results of site visits can include delight, questions, complaints, interference with work, theft, injury, and even buyer installed items (speaker wire, for instance). Understandably, builders hope to avoid all but the delight scenario.

Your challenge is to balance control of operations with empathy for buyers' need to watch the home being built. The task is further complicated by the reality that enforcement of site visit policies falls somewhere between somewhat difficult and utterly impossible. To address this issue you need site visits policies, a procedure for managing buyer questions and lists during construction, proactive and frequent contact with buyers, and enforcement steps for times when enthusiastic buyers overlook your policies.

FIGURE 8.1 Frame Stage Meeting Agenda

<Builder Logo>

Frame Stage Tour

Name	_____	Date _____
Address	_____	Community _____
Phone (H)	_____	Lot# _____
Phone (W)	_____	Plan _____
Fax	_____	Target Closing _____

A tour was completed on thes date to review the topics listed below and confirm corerct installation of selections visible in the home at this time. Some items listed on sleection sheets or described in change orders may not be apparent at this stage of construction.

Exterior

* Homeowner manual
* Elevation
* Exterior finish materials
* Meter locations
* Air conditioner condenser
* Patio/deck
* Hose bib locations
* Property bouncaries
* Drainage swales
* Driveway
* Sidewalk
* Fence installation

Interior

* Foundation
* Beams and supports
* Framing options
* Doors and windows
* Ceiling details
* Engineered components
* Trusses
* Roof sheating
* Flashing
* House wrap
* Electrical options
* HVAC options
* Plumbing options
* Next step: Orientation

Follow Up Items

_____
_____
_____
_____
_____
_____
_____
_____
_____
_____
_____
_____
_____
_____
_____
_____
_____
_____
_____
_____
_____
_____
_____
_____
_____
_____
_____
_____

Purchaser	_____	Date _____
Purchaser	_____	Date _____
[Builder]	_____	Date _____

Site Visit Policies

Attempting to keep buyers away completely may make buyers wonder "What are they hiding?" Accepting that buyer site visits are likely, under what circumstances are those visits acceptable? Begin this decision making process by thinking through practical issues.

Staff-Accompanied Visits. Do you hope to have buyers make appointments with a staff member and be accompanied on-site? If so, which staff members will be available for site visit duty, and when? An advantage with this method is that a trained staff member handles the buyers' questions. Safety procedures such as wearing hard hats are more readily enforced and most home buyers appreciate the attention. However, these advantages come at a price.

How will site visit duty affect the other job responsibilities for these staff members? The sales person who accompanies buyers on-site is unavailable to new prospects and is not making follow up contacts. Likewise, a superintendent who is touring the site with buyers is not scheduling trades, walking homes to check quality, or ordering materials.

Personnel with less critical job assignments might be enlisted to accompany buyers-a sales hostess or an assistant superintendent. Do these employees have the necessary knowledge, communication skills, and authority to respond to questions?

Issues raised during random site visits should be documented. A generic form like the one in Figure 8.2 can accomplish this and carry through the same formatting as planned meeting agendas. Follow up where needed should be just as rigorous as for standard meetings.

Preferred Site Visit Times. Whether you ask that buyers be accompanied or not, are certain days of the week or certain times of the day better for buyer visits than others? Requesting that buyers visit outside of the normal trade contractor work day can reduce interference with on-site labor. Buyers are more likely to ask questions of the company personnel you have directed them to if trades are unavailable.

Standing Appointment. Small volume companies get good results from a planned site visit, usually on a weekly or bi-weekly basis, depending on the stage of construction. The home buyers and the superintendent know they will have each other's undivided attention at this routine appointment.

Insurance. Read your insurance policy and if necessary talk with your agent to understand what protection you have for customers on site-and what your liabilities are. A construction site is an attractive nuisance-meaning it draws curious people who may combine exploration with injury or mischief. In most cases, the builder (and supporting insurance company) often suffers the loss if any problems occur.

FIGURE 8.2 Random On-Site Meeting

<Builder Logo>

Site Tour

Name	_____	Date	_____
Address	_____	Community	_____
Phone (H)	_____	Lot#	_____
Phone (W)	_____	Plan	_____
Fax	_____	Target Closing	_____

Follow Up Items

_____
_____
_____
_____
_____
_____
_____
_____
_____
_____
_____
_____
_____
_____
_____
_____
_____
_____
_____
_____
_____
_____
_____
_____
_____
_____
_____

Purchaser	_____	Date	_____
Purchaser	_____	Date	_____
[Builder]	_____	Date	_____

Buyer Concern System

Whatever your site visit policy you decide upon, standardize your method for tracking and responding to buyer questions during construction. This means documenting the questions and the answers. Providing buyers with forms to note their concerns shows that questions are a normal part of the process. Strive to *acknowledge* buyers' questions the same day you receive them and to *answer* the questions within 48 hours.

By noting the date and time a question was received and when the company responded, you can measure response time-a critical element in buyer satisfaction. Routine review of the buyer concerns may identify common issues that you could discuss up front with all buyers-often these subjects fit best in the preconstruction conference. The completed form then goes into the file as part of permanent records and may be helpful to other personnel who work with the buyer. Whatever system you decide upon, make certain that it consistently produces the promised results, otherwise your home buyers will invent their own system to get what they need.

Buyers' Lists

Customers often bring long, detailed lists into the sales or construction office. These frequently include instructions to install windows, carpet, or shingles. After receiving such a list, many are tempted to respond, "We probably would have thought of that!"

Tossing the lists into a desk draw and ignoring them can be embarrassing, expensive, humiliating–or all three. Hidden among such suggestions as "connect electrical service," you may find things of which you were unaware such as the fireplace is in the wrong location or a blue bathtub was delivered even though an almond one was ordered. Hearing an "I told you so" from a frustrated customer is an unpleasant experience and should be avoided whenever possible.

When a buyer presents a list of concerns, acknowledge it immediately and respectfully. Although you may think your company owns the home, this is only true financially and legally. The buyers own it psychologically and emotionally. Condescending reactions are guaranteed to generate dissatisfaction. Besides, you may be glad the buyer pointed out some of the details listed. Review each item and respond within 24 to 48 hours if possible, just as you would for a single question or concern. If all this seems like a lot of bother over little details, remind yourself that you're getting a lot of little dollars for the effort.

Routine Contact

Many companies have discovered that initiating regular communication with buyers has a positive impact on their satisfaction while reducing random site visits. Customers are less likely to interfere in the work if they understand what is going on and have real trust in the company.

Some larger volume companies have created a position designed to address buyers' needs during the building process. Carrying titles such as Home Buyer Liaison or Home Buyer Service Manager, this individual takes over after the purchase agreement is signed, oversees all steps in the process, and communicates with the buyers up through closing.

Mid- to small-volume companies have discovered that tracking and confirming a minimum of weekly communication with each buyer is worthwhile. The sales person or the superintendent-often following the weekly on-site meeting, can initiate the contact. The update is done by in person, by phone, fax, or email.

Educate Buyers on Site Visit Policies

Once you have clearly defined your site visit policies, you can plan the steps and create the materials needed to convey them to your home buyers–the teachable moment. Builders have several opportunities to present site visit policies to buyers before construction even begins–in the contract, the homeowner manual, preconstruction conference. Use all of them. Then be prepared to respond if home buyers overlook your policy and interferes with work.

Purchase Agreement. An example of a noninterference clause is offered in the fourth edition of David Jaffe's book, *Contracts and Liability* published by Builder-Books. As always with any legal matter, be sure to check with your attorney for specific requirements of your state. This clause was adapted for use in the letter that appears in Figure 8.3.

Homeowner Manual. Follow up the purchase agreement clause with a description of your site visits policies in your homeowner manual. Forms documenting questions give your on-site team a method for tracking and get the home buyer into the habit of documenting items.

Preconstruction Conference. Include site visit policies on the preconstruction meeting agenda. The superintendent should refer to the homeowner manual, call attention to your buyer concern forms, and discuss the benefits of their use for all concerned. Now is the time to align the buyers' expectations regarding the response time for their questions. Tell them that most questions can typically be answered within 48 hours and if more time is needed, you will advise them of that.

Enforcing Site Visit Policies

With all this up front effort, most on-site problems are avoided. However, when interest becomes interference, you need to react quickly to prevent things from spinning completely out of control. The superintendent can begin with a casual in-person chat or a phone call about this issue.

The second time that buyers interfere with work, firmer measures may be needed. The letter in Figure 8.3 addresses buyer interference with understanding and firmness.

FIGURE 8.3 Letter: Buyer Interference

Dear <Home Buyer>:

Your interest in your new home is understandable. We appreciate that you want a quality product and value for your money. Our construction systems are designed to provide you with the quality and value we promised you in our contract documents and showed to you with our model homes.

To achieve this, it is imperative that [Builder] direct and maintain control over the work. This is such an important issue that the [Builder] purchase agreement which you signed on <date> includes a clause which states:

Purchasers shall not in any manner interfere with work on the job nor with any subcontractor or workers. Purchasers will not communicate directly with the builder's workers, employees, agents, or subcontractors regarding the means, method, or manner in which they are to perform the work. If Purchasers delay the progress of the work, causing loss to the builder, the builder shall be entitled to reimbursement from the owner for such loss.

On several occasions conversations you have initiated with trades people and [Builder] employees have caused disruption of the work and confusion about methods and materials. We have asked that you communicate any questions you have only to your sales associate. [Builder] has a written system for tracking and responding to such concerns.

We ask that you respect this system and refrain from any further interference with work at the job site. [Builder] will pursue all avenues of recourse to recoup any losses caused by further interference.

We appreciate your cooperation in this matter. If you have any questions, please contact me.

Sincerely,

<Builder>

If this more formal communication fails to stop the home buyer's interference with work, you are left with two choices. The first is to put up with the buyers "helping" you build the home. Backing down on clearly stated policies sends a clear message that your policies can be ignored or overridden-count on future problems. The second alternative is to threaten cancellation of the contract. With the documentation you have created along the way, you should have little difficulty establishing that your home buyer is in default of your contract.

While threatening to void a contract is certainly undesirable, long term it may be the wisest action. The threat may stop the buyers' inappropriate behavior; if it does not, canceling the contract may be your last resort. Avoid such regrettable circumstances by communicating clearly and earning the buyers' trust from the start.

Change in Product

Most purchase agreements say the builder has the right to "substitute equal or better materials or methods." However, genuine concern for home buyer satisfaction means making a reasonable effort to alert buyers when you make significant changes. The question of whether a change is significant is a matter of judgment. If you error in judging this, your buyers will call the oversight to your attention. Often, told in advance–see Figure 8.4–to expect a difference, they not only accept the change without complaint, but are impressed with your attention and forthright communication.

Delivery Date Update

Permitting, weather, labor shortages, material delays, inspector schedules, buyer change orders . . . many events and circumstances make accurate deliver date predictions a challenge for today's builder, especially during the early stages of construction.

Some buyers handle this ambiguity better than others. All buyers expect and appreciate seeing their builder manage work on their home effectively. While providing a definite delivery date may be impossible at the beginning, nothing should prevent a builder from keeping home buyers informed of the status of their homes.

Large corporations can afford the production costs for special cards, perhaps with color graphics, to update buyers that key points of construction have been reached. Smaller operations may not have the resources for elaborate printing, but any company can send a letter like the one in Figure 8.5. The effort goes a

---

FIGURE 8.4 Target Delivery Date Update

Dear <Home Buyer>:

As you know, [Builder] reviews construction schedules on a weekly basis. We have identified several benchmarks in the construction process that help us determine first a target and later a confirmed closing date. We know you are vitally interested in that information and have many details to arrange around the closing date.

Your home reached is now at <stage>. The current target closing date is between <date> and <date>. This date may move as construction progresses.

When the home reaches cabinet stage, we will set a firm closing date, giving you a minimum of 30 days notice. Until we set a firm closing date, we recommend that you keep all financial and moving arrangements tentative.

We will be happy to discuss any particular details you may need information about. Please contact me if you have any questions.

Sincerely,

<Your Name>

FIGURE 8.5 Letter: Change in Product

Dear <Home Buyer>:

The specifications for your home describe the water heater as a <brand, model, size>.

The manufacturer ceased production of this model and it is now unavailable for your home. In its place we will install a <brand, model, size>. We reviewed features, operational costs, and manufacturer warranties from over 10 different choices and selected this as the most comparable to the original.

We believe you will be satisfied with its performance.

If you have any questions, please contact me.

Sincerely,

<Builder>

long way toward keeping customer confidence and is especially valuable in regions where trade shortages makes the construction schedule take longer than anyone involved wants (an area roughly defined as East coast to West coast and from Mexico to Canada).

Chapter 9

Orientation

Your company stands to gain tremendous advantages from vigorous management of the home delivery process. Consider, for example, the following:

- For days before, during, and after their move, home buyers talk of little else. During these conversations, your company's name is likely to come up. These clients are walking-talking billboards for your company.
- When homeowners move in believing that they were treated well and received what they paid for, they make more reasonable requests of warranty. Angry homeowners, on the other hand, submit revenge lists and are difficult to work with every step of the way.
- Human beings can form a habit in as little as 21 days. After months of behaving in buyer mode, many clients have difficulty changing buyer behaviors to owner behaviors. A ritual or ceremony that marks the change from buyer to owner helps your home buyers make this transition.
- Denying action on excessive customer demands is easier if the home is truly ready to deliver. A home in poor condition stimulates aggressive attitudes with many buyers. At the same time, you will feel a pressure to give extras in an effort to compensate the home buyers for your company's failure to perform.

To create an effective orientation program begin by finishing homes for delivery.

Finish the House

The first impression sets the tone for the orientation and often for the warranty period. To make this first impression a good one, begin with a detailed definition of a complete home and a firm commitment to meeting those criteria. Then institute procedures to reach that goal consistently.

Final Punch List

Pragmatic superintendents contend that sending tradespeople through the home twice—once for the punchlist and again for the orientation list—duplicates work. However, if the punchlist is thorough and the trades attend to the items noted, the home makes a good first impression and any follow up is minimal. When punchlist items remain incomplete, buyers understandably become concerned and perhaps even angry-leading to more and more items. Buyers should not need to create a punch list on their new homes: Quality control is the builder's job.

Quality Assurance Inspection

A further step, following completion of the superintendent's punch list, is a quality assurance (QA) inspection performed by the superintendent and a quality assurance or warranty rep. This joint inspection is intended to be a confirmation that the home is ready to deliver, not an opportunity for the quality assurance or warranty rep to help create the final punch list. To work well this requires that construction personnel have time (usually one to two days) to correct noted items.

Prep Team

Prep teams include trained personnel who work from a standardized list to detail the homes for delivery. Feedback from orientation reports helps them fine tune their list of items. Establish criteria for the condition of homes turned over to the prep team prevents their needing to do inappropriate amounts of finish work.

Zero Defect

Many builders have instituted a zero defect program and advertised that decision to home buyers. Experience has shown this creates more problems than it solves. After months of promoting the new program, many of the builders who applied it are now dropping the term from their home buyer literature and conversations. And with good reason.

Most buyers misunderstand the meaning. Originally a military concept, quality expert Phil Crosby popularized the term zero defect in his writings and teachings. The term refers to the standards set by the organization and signifies a commitment to produce products with every component meeting those standards. Unless they interpret "zero defect" the same way as the builder, home buyers can easily conclude this promise means the home will meet their personal definition of perfection. As this in not an expectation most builders can live up to, the term is best left to the military.

Who Should Conduct Orientations?

This debate rages on–superintendent, warranty rep, or a full-time orientation rep? The answer may be a function of company size and the number of personnel available, however, in the event that you have several options, the pros and cons of each are listed here.

Superintendent

Certainly the superintendent who built the home knows it best and is familiar with issues the buyers raised during construction. Also, by writing the list personally, the superintendent understands what corrections are needed. Many companies have taken the position that if the superintendent has to face the home buyer he or she will learn to get the homes complete. Two problems exist with this approach. First, the tool for teaching the superintendent is that the home buyers must get upset-an event wise builders work to avoid. Second, the superintendent's habits seldom improve. Instead, many superintendents complain about their "customers from hell" and fail to see their role in creating that hell. This procedure further assumes the problem with finishing homes is entirely within the superintendent's control. If this method worked, all of the companies using it would be delivering complete and clean homes–most do not.

Phantom lists ("We don't need to write that down. I already knew about it.") and overflow to warranty ("Don't be concerned about that right now, just put it on your first warranty list.") are common when the superintendent conducts the orientation. Time spent conducting orientations is time not spent building homes. Homeowners may continue to report items to the superintendent after move-in, circumventing the normal warranty documentation system and increasing superintendent workload. Performing orientations means the superintendent needs a full range of customer relations skills. Some are unable to develop these skills; others may not want to.

Warranty Rep

In an approach typical of mid-size companies, buyers meet the warranty rep at orientation. This often starts the warranty period off with a positive rapport. Customers work with a warranty rep focused on developing customer relations skills.

Inspection is likely to be thorough, yet reasonable. Overlooking items results in long warranty lists later. On the other hand, being too nit-picky teaches the buyers to examine their home with that same attitude during warranty. Buyers benefit from warranty rep's awareness of past problems in other homes. Later cosmetic claims are easier to resolve when the same person inspects the home at orientation and during warranty.

Disadvantages of this approach include the reduced time the warranty rep has for warranty work. Scheduling orientations around the warranty rep's appointments with homeowners can be difficult. Friction can develop between the superintendent and the warranty rep over the items noted. The superintendent may come to rely on the warranty rep to prepare the final punch list in the guise of an orientation list. Finally, the warranty rep may be unfamiliar with buyers' selections, change orders, or issues the buyers mentioned during construction.

Orientation Rep

Suitable for large-volume companies, this approach allows one or more staff members to work full time on orientations. Consistency is likely in the presentation and standards used to judge the homes. Meanwhile, superintendents are building homes and warranty reps focus on warranty. However, even these specialists will not prevent buyer dissatisfaction if the homes are incomplete. In slow times, paying orientation reps when there are few orientations can be painful.

Third Party Inspector

With this approach, which can be used by companies of any size, buyers receive third party objectivity. Inspectors who see homes by other builders have a broad perspective on quality in the region. Superintendents are free to focus on building and warranty reps focus on warranty. The builder pays only for the services used.

However, appointment times may be limited by other work the inspection company has scheduled. The inspector may be knowledgeable but lack good people skills. Turnover within the inspection company can result in unknown people presenting the homes. Turnover also makes it difficult to keep the inspector familiar with your product and policies. Finally, the per hour cost is usually higher than with the other strategies since the inspection firm is in business to make a profit also.

Regardless of which method you choose, homes must be complete and clean for any program to succeed. Identify the best method for your company, keeping in mind the desirability of a clean break between construction and warranty. A clean break means that all involved staff recognize where construction's responsibility ends and warranty's responsibility begins.

Schedule the Orientation

Orientations require day light and should begin no later than 3:00 p.m. in most regions. To make efficient use of personnel, some builders designate certain days of the week for appointments in specified subdivisions. Another aspect of scheduling is how far in advance of the closing should the orientation occur? Field personnel need time to react to the orientation list. Three days should be adequate to complete or significantly reduce the list and deliver a finished home. Only items that required ordering parts should remain.

Provide your buyers with scheduling information in your homeowner manual. When setting the appointment with the home buyers, review key points such as:

- The purposes are to demonstrate the home and confirm correct completion.
- The orientation will take approximately two hours.
- The agenda includes several hundred details; please attend alone to focus attention on the information presented.
- You can find more information and copies of the orientation forms in your homeowner manual.
- Review "Caring for Your Home" section of your homeowner manual and note any questions.
- You will be touring the exterior, wear suitable shoes. We provide paper "booties" to cover them when we go inside.

- A last minute flurry of activity is normal; many people will work in the home during the last several days leading up to this meeting.

Presentation

Well-done orientations look easy. When a competent orientation representative presents a new home to clients, the meeting flows smoothly. Such a meeting includes a bit of theater, a wealth of useful information, and a fair examination of the home. Orientations work best when their focus is on education and celebration.

Training

"Here is your day/night thermostat. The book that explains it is in the kitchen drawer," is not an orientation. Acquiring the knowledge required to do orientations effectively is a challenge that necessitates genuine study of products and methods. Detailed knowledge about the community, construction techniques, and the products in the home is essential. For components such as the furnace or water heater the discussion needs to be logical. Keep operating steps in sequence. Jumping around is confusing, takes longer, and risks that key points will be omitted.

Itinerary

The itinerary is the route around and through the home. Without a strategic itinerary, the route taken through the home can be haphazard and varies from home to home. The advantages of a consistent itinerary are worth the effort:

- When orientations follow the same itinerary, the possibility of overlooking anything is reduced.
- A standard pattern and time table evolves; you know how long each segment should take and can gauge whether the discussion is off track.
- Technicians who work in the home following the orientation can more easily find items when they are listed in a standard order.
- Basing the orientation on a well-considered itinerary allows you to present the home in the best possible order and celebrate the event with the buyers.

Begin at the Street. An itinerary that produces excellent results begins with greeting the buyers at the street. As they get out of their vehicle, welcome the buyers, exchange the usual amenities, and then pause to look at the house. Let the buyers savor this moment. "Congratulations, Mr. and Mrs. Jones, here's your new home." (Note: garage overhead doors should be closed.)

Avoid being in such a hurry that this gratifying moment is lost. Simply looking at the home for these few seconds accomplishes several important things. First, the moment is gratifying to buyers who anxiously waited for the home to be completed. Second, it provides a healthy perspective for the rest of the meeting. Viewing the home from the street fixes the size of it in the home buyers' minds. If you note a few minor items for attention, they are seen in perspective–against the image of the entire home. And third, this ceremonial moment also begins the vital transition from buying to owning.

Exterior First. The moment at the street leads naturally into starting your presentation there. As you discuss the home's exterior, follow a set pattern, such as going clockwise. The emphasis is education rather than inspection: help the your clients understand how each component of their home works, how to care for it, and what the builder's limited warranty commitments are.

Front Entrance. After completing a tour of the exterior, return to the front of the home. Enter the home through the front door–as if the home buyers are guests. Exteriors are not as finely detailed or tightly finished as interiors. After looking at the outside for 15 to 20 minutes, the sparkling interior looks even better.

Demonstrate and discuss the entry features and then, depending on the geography of the floor plan, visit each room of the home, ending in the kitchen.

Kitchen Last. Demonstrate the kitchen last. You will arrive with the buyers' confidence high and anxiety low. This brings us to one more reason to update the traditional kitchen-first itinerary. Another negative associated with starting in the kitchen is the paranoia-inducing inspection for cosmetic damage.

Nothing in the updated itinerary is intended to prevent noting legitimate items, whether cosmetic or functional. In fact, you should volunteer to note items that fail to meet your promised standards, including cosmetic items. After all, your company's name is on this home.

Notice the difference between this approach and the itinerary that takes buyers directly to the kitchen, where traditionally the orientation began with a review of the manufacturers' materials about appliances. However, starting with uninteresting paperwork and progressing to nervous inspection of cosmetic surfaces turns what should be a celebration into boring work at best and a nerve wracking challenge at worst.

As if that's not bad enough, the buyers' habit (which the traditional itinerary helps to create) of looking at every kitchen surface from a distance of three inches continues through the rest of the home. This leads to either denials of requested items or a long list of minute items that will have technicians rolling their eyes.

End the tour of the home in the kitchen with an orderly demonstration of kitchen features. Show how appliances work and discuss care of surfaces. Use the counter space to review paperwork, briefly reviewing the manufacturer's literature. To maintain a positive tone, when discussing warranty coverage begin with examples of concerns the builder will repair. Then give examples of homeowner responsibilities. Thank the customer and walk out with them. Procedures describing these steps appear in Figure 9.1.

Closing Day Inspection

An optional step is to schedule a brief closing day visit to the home to confirm completion of items and reassure buyers that no damage occurred during that

---

FIGURE 9.1 Procedure: Orientation Presentation

Objectives

Home Buyers

1. Maintain a high level of comfort and satisfaction with their new home.
2. Are satisfied that the home delivered meets or exceeds the standards shown in the show homes.
3. Understand what to expect from the home as time passes, including the effects of expansion and contraction, shrinkage, weather, and normal wear and tear.
4. Understand their responsibilities for maintenance of their new home.
5. Confirm the acceptable condition of cosmetic surfaces subject to items listed on the orientation form.
6. Understand and accept [Builder]'s procedures and time frame for completing items.
7. Know the procedures for obtaining routine or emergency warranty service.

Procedures

Orientation Rep

1. Check with the salesperson and the superintendent for any last minute information.
2. Arrive at the home 20 minutes prior to the appointment.
3. Review the home and prepares for the orientation.
   a. Check and adjust temperature.
   b. Turn on lights.
   c. Close closet doors; lowers garage overhead doors.
   d. Inventory and assemble manufacturer literature.
   e. Note any items that fail to meet [Builder]'s standards.
4. Greet the buyers at the street and welcomes them.
5. Review the purposes of the meeting and introduces the procedure.
6. Conduct the orientation according to the [Builder]'s itinerary and agenda. (attach your orientation training agenda.)
7. Obtain the buyers' signatures on the orientation forms and provides the buyers with copies.

(Continued)

FIGURE 9.1 Procedure: Orientation Presentation (Continued)

8. Conclude the orientation with review of what has been accomplished and procedures for warranty visits or warranty emergency.

9. Immediately deliver the list of items to the designated person.

Materials

1. Orientation forms

2. Homeowner manual

Assessment

1. Customer satisfaction surveys receive positive ratings for orientation questions.

2. Few questions or requests for non-emergency warranty service are received by warranty between the orientation and first warranty visit.

3. Few maintenance items are listed on warranty service requests.

\<revised date\>

work. Though time consuming for all involved, this can work acceptably if the items are actually completed. Re-visiting a home to note that only 3 of 28 listed items have been addressed is disrespectful of buyers' time and calls more attention to the poor service.

A frequent concern with this confirming visit is that the home buyers may want to create a new list of items. Choices in response to this predictable scenario are to make another list and go through the cycle again or hold the items for the first warranty visit. Neither is desirable. Finish the home for the first presentation and all of this extra effort becomes unnecessary. Doing so can also prevent arguments over escrows.

Escrows for Incomplete Items

When incomplete items exist, some buyers (or their lender) may demand that a portion of their final payment be placed in escrow until all items are complete. If none of the other good reasons for delivering complete homes motivates your company to complete homes, this one should.

Postponing the closing until the items are complete may be a better long-term choice than agreeing to an escrow. Escrowed funds can easily lead to a power struggle. The original list of items may be complete but the homeowners refuse to release the escrow until a new list is finished. If escrowing funds is unavoidable, your best protection is to have an objective third party confirm completion of items and release the balance.

Orientation Closure

Perhaps the worst kept secret in the new home building industry is that no one really wants to do orientation items. Interestingly, when buyers refuse to close until all items are completed, those items are typically completed within a day or two. Clearly, this is a largely matter of motivation and commitment. When management makes orientation items a priority, results follow. Set a firm time frame for completing these items. Document this requirement trade contracts. Make it clear to staff and trades that completing orientation items means completing all of them–not nine out of eleven, but all of them. This is an attainable goal. Builders who achieve it are easily identified by their euphoric smiles.

Prompt Response

Identify the trades you typically need for orientation items and keep them informed of upcoming orientations. Communicate items to the appropriate people immediately and accurately. By copying the original list and highlighting or circling those items for each trade on its copy, everyone will have the same information. Time spent rewriting or typing the list is time wasted; a legible list eliminates this extra work. Keep the original list clean and in a safe place.

As trades complete their items, they should sign and date their copy of the list and return it. The repairs performed should be correct and effective; band aid approaches impress few homeowners.

Homeowner Acknowledgment

A common builder complaint is "We pay a $50 per house bonus for a signed off orientation list. Yet many items from those lists appear on subsequent warranty requests. Why do customers sign the list if the work is not done?" Because the superintendent gets them to sign it. "I've talked to all the trades people we need and they're scheduled to do this work. I need to get the paperwork turned in," or "Just to get the paperwork out of the way, why don't we go ahead and sign it . . . ." Good intentions notwithstanding, signed lists receive little attention. Once the honeymoon ends at about 30 days, the homeowner complaints start. Avoid this with effective checks and balances.

Follow-up with the new homeowner serves several purposes. The customer appreciates the attention. By acknowledging that orientation items have been completed, the customer consciously recognizes the good service. This is best done with an in-person visit, although phone, email, or traditional mail can be used. If the homeowner has any complaints about the work performed, the work in question should be inspected and a determination made as to what further attention-if any-will be provided. Once all items are completed to within company standards the homeowner and the superintendent should sign and date the original list for the file.

Closing Date Confirmation

From the beginning home buyers are understandably curious about when they can move into their new home. Most builders find it difficult to predict that date with certainty because of many factors outside the builder's control. At some point, however, the home is far enough along that fewer factors can impose delays and the builder should commit to a firm delivery date and notify the buyers. The more notice buyers receive for their closing date the better, ideally a minimum of 45 days, though many companies can give only 30. For buyers who want to lock a loan, close on the sale of their current home, schedule movers, give notice on a rental they occupy, and arrange other details, anything less than 30 days can cause them expense and inconvenience–which they understandably see as the builder's fault. Avoid this if at all possible.

The closing appointment is usually arranged with the buyers by phone and that conversation should include a review of key preparation steps with the buyers. Confirm the details in a letter such as the one in Figure 9.2.

Transition Service

The excitement of the process and momentum build until the home is ready to deliver and an orientation is scheduled, followed by a blur of paper at the closing and the exhausting work of moving. Suddenly the excitement is over and the buyers are left with the mundane tasks of hanging pictures and getting rid of empty boxes. By planning several proactive contacts shortly after closing, the builder can wean the homeowner off the excitement and generate benefits for the homeowner and the company.

Plan a minimum of three contacts during the first month, perhaps including one in writing, one by phone, and one in person. One contact might take the form of a thank you letter from the company owner. The superintendent might stop by in person to check on orientation items. Sales might stop over to give the new homeowner a thank you gift. Consider giving your home buyers a supply of "We've moved to our new [Builder] home" note cards with space for the new address and phone number inside. This provides buyers with a genuine convenience as they update friends and relatives while it advertising your company at the same time. The first warranty visit can also be part of transitional service–details appear in Chapter 10.

FIGURE 9.2 Letter: Closing Appointment Confirmation

<Date>
Dear [Home Buyers]:

This is to confirm your closing appointment which has been scheduled for—

<date> at <time>
at <address>
If you have any questions, please contact <name> at <phone>.

The asterisk on the enclosed map indicates the location for your closing appointment. Parking is available on the north side of the building or two-hour metered parking can often be found along the street.

The closing process usually takes approximately one hour. In preparing for this important meeting, please include the following items:

- Transfer utility services. Phone numbers are listed in your homeowner manual in the sections on closing.
- Arrange for evidence of insurance. Your insurance agent will know what is needed. Allow two weeks for this.
- Plan to bring certified funds. The exact amount is usually calculated very near to the closing date since some items are prorated to that day.
- Confirm with your mortgage lender that all loan contingencies are satisfied. If any further documentation is required, be sure to bring that to the closing.
- Closing agents have no authority to negotiate for lenders or builders. If you have any remaining questions, work directly with your lender or contact this office to obtain needed answers prior to the closing.

Soon you will be moving into your new home. We look forward to having you join our community.

Sincerely,

<Builder>

Chapter 10

Warranty

Homeowners prefer that the components in their new home perform properly. Their distant second choice is to have prompt, courteous, and effective warranty repairs (translation: on-time, quality work, including clean up of the work area, and excluding collateral damage). In other words, they want results. Handling warranty items quickly and smoothly is one secret to maintaining homeowner goodwill and keeping costs under control. Warranty personnel and homeowners should clearly understand what services you provide and how procedures work. A good warranty service program leaves as few questions as possible.

Warranty Service Structure

Commit to a level of attention that you can provide consistently and effectively. Satisfaction is greater when builders promise a bit less and do it well than when they promise a lot and deliver only half of it. A good balance comes with a 30- or 60-day visit and minimally a year-end letter.

Builder-Initiated Warranty Visits

For years builders told homeowners to send in a warranty list 30 days after moving in causing homeowners comment that once the builder gets paid, interest in customer satisfaction disappears. Increases in customer expectations for attention and service have influenced that procedure. Service oriented builders ask to inspect the home to confirm that all components are functioning correctly. The procedure involves three simple steps:

- Develop a warranty meeting checklist similar to the one in Figure 5.7. This checklist helps to establish the warranty rep as an authority on company standards, making it easier to deny excessive requests.
- At the end of the orientation, ask the home buyers to set an appointment for their first warranty visit, suggesting that it be set for 30 to 60 days in the future. Knowing they have an appointment (even if it is changed later) results in fewer calls to the warranty during the first weeks in their new home.
- Several days prior to the appointment, the warranty administrator contacts the homeowner to confirm the appointment. Ask whether the homeowners have noted any warranty items and if so, request that they fax or e-mail the list to the warranty office. Builders using this approach report less warranty work.

Year-End List

Builders who use year-end reminders report that the volume of year-end work remains about the same. Arguments with homeowners who forget their warranty expiration and want to report items several months later drop dramatically. By controlling the timing of year-end warranty visits, you will have fewer disagreements with trades about their warranty liability. You can also confirm that your homeowners are maintaining drainage properly and answer maintenance questions.

Emergencies

Accept reports of emergency items by phone and respond immediately. Your definition of emergency items and associated procedures should appear in your homeowner manual.

Miscellaneous Warranty Requests

Homeowners will occasionally report items between the standard checkpoints. Remind them that they have the year end warranty service appointment coming. If they persist, proceed cheerfully with appropriate attention. Inspect and process such items according to the same procedures and standards as a 60-day or 11-month list.

Cosmetic Damage Noticed after Orientation

Nearly every builder's orientation form includes a statement releasing the builder from cosmetic repair responsibility. Simple enough in theory, but the reality is much more complicated. At the orientation, typically three or more adults tour a clean, empty home in broad day light, with the purpose–among others–of confirming the good condition of cosmetic surfaces Although the likelihood of overlooking significant surface damage is low, it can happen. More common is cosmetic damage that appears after homeowners move in without their having caused it. Minimally, you owe the homeowners the courtesy of an inspection.

Consider this example. If the electrician drops a tool into the tub during construction and is relieved to see no damage occurred, he may be leaving behind a crack that is invisible to the eye. When the homeowner uses the tub, the weight and temperature of the water causes the chip to appear. Damage occurred after move but the homeowner did not cause it.

Establishing "back door standards" can help in such situations. These guidelines are not published to homeowners, but provide boundaries for discretionary decisions. You might develop yours based on a combination of a number of days, an amount of money, or specific repairs. For instance, your back door standards might provide your staff with the latitude to:

* Repair cosmetic damage discovered during the first five days.
* Spend up to $100 for cosmetic repairs at any time.

- Replace any window that gets a stress crack.
- Repair a countertop, resilient floor, tub, or sink. If the homeowner insists on a replacement, provide the appropriate names and numbers for them to arrange and pay for this on their own.

"Out of Warranty" Warranty Requests

Builders sometimes believe that when the clock strikes midnight on the last day of a homeowner's material and workmanship warranty, the company's obligation ends. Actually many conditions can result in a builder correcting items after the warranty expiration date:

- Grace periods, which vary from 10 to 30 days, provide a common sense time period for a homeowner to notify the builder of items noted right at the end of the warranty.
- Builders must comply with the codes that were applicable at the time of construction, regardless of the status of the warranty.
- Builders must fulfill change orders and selection sheets and provide advertised standard features in the home even if the homeowner does not notice the omission until the warranty has expired.
- A latent defect is one that could not be discovered through normal inspection but existed from the beginning. For example, incorrectly installed valley flashing allowed a roof leak and damage appeared when the home was three years old. The original flashing error that allowed this was not discernible in a normal inspection. That the warranty had expired did not release the builder from repair obligations.
- If the homeowner reported an item in writing during the warranty period, but closure was not documented, the builder must respond even if the warranty has expired.
- If the same problem was repaired twice or more during the warranty, the failure to repair it satisfactorily might subject the builder to a breach of warranty claim. For instance, if the air conditioner misbehaves the same way it did twice during the warranty period, the builder's obligation (as well as the installer's and manufacturer's) continues.
- Consumer product warranties often provide protection for the homeowner beyond the builder's coverage. The homeowner may require assistance from the builder in these matters, especially since the builder will have more clout with the manufacturer. Provide this support to help your customer and to learn how the manufacturer stands behind its products.

You may encounter other circumstances that justify repairs outside the warranty period. Remain objective, listen carefully, and investigate fully as you would for a warranty request during the warranty period.

After Hours Service

Builders have several choices for managing after-hours service-none of them perfect. Each possibility comes with advantages and disadvantages.

Emergency Numbers. These are typically delivered at the orientation to ensure that the company names and numbers are those of the trades who actually worked on the home. Many builders use this method with good results. While cost is minimal, this fairly common approach leaves the homeowners and trades on their own until the builder's next regular business day. Although trade response can be excellent, it can also be otherwise. And builders do not always find out about the incident or get it documented for the home's warranty file.

Answering Service. If you take this approach, be certain you understand fees, what services are available, how much scheduling flexibility you have, staff stability, and the reliability of the answering service company's technology. To prevent confusion, assign one person from your company to handle communications with the answering service firm.

On-Call Duty. This method saves the cost of an answering service but long term, staff may show the effects-especially if off-hours calls are frequent or angry. An often overlooked concern is the pressure family members may put on the "on-callee" because of interruptions in family time. To institute a reasonable on-call system, ask first for volunteers for this duty. Consider geography. On-call personnel may need to visit the home and if that means a two-hour drive, resistance is likely. The person on call needs construction knowledge, good communication skills, and the authority to direct trade contractors. Set up the schedule well in advance with maximum weekly increments. Expect some trading of on-call duty as family needs arise. Perhaps most importantly, establish some compensation-time off, a monetary reward, or some privilege (one company owner gave "on-call" staff weekend use of his ski condo once each winter as a thank you-the employee only had to pay taxes on the value).

Response Time

Good service is fast. Benchmark your service times against these standards. If you performance misses these targets significantly, that may mean you have too few staff or too much warranty work. Acknowledge requests for service within one business day. In that communication, offer an inspection appointment time within the next five business days, although the customer may be unable to take that appointment due to his or her schedule. Issue work orders resulting from inspections within one business day of the inspection. Return phone calls as soon as possible, but within four hours as a maximum.

Traditionally builders used a 30-day time frame for completing repairs on warranty items. As the focus on customer satisfaction has increased and other businesses have accomplished faster response times, builders have become more aggressive in this area. Ten work days is now the goal of choice. Success is achieved when you stay within that time frame with 90 percent of warranty repairs.

Rigorous Monitoring. Frequent and consistent monitoring are at the heart of effective warranty service. Follow up on work orders before they reach their expiration date. Minimally once a week, notice which work orders remain incomplete and do something about them. If you wait until monthly reports are compiled, circulated, reviewed, and acted upon, the homeowners may be picketing before anything improves. Sending an unpleasant note to a trade who has fourteen 65-day-old service orders is taking too little action too late.

Follow Up with Homeowners. Follow up contact can put a builder's reputation in a class by itself. In as many cases as possible, contact the homeowners-by phone, fax, email, post card or letter, or in person visit–to confirm that the problem was solved. Especially on the more significant issues this contact is essential. When trades know you make follow up contact, they are often more thorough in providing repairs. Even if you "know" work is complete, the contact impresses the homeowners. If the homeowner is dissatisfied, the sooner you know the better.

Warranty Reports

Warranty work lends itself to many types of reports. Begin with the basics: number, nature, completion, and cost of warranty work.

Number of Warranty Items

Calculate the number of warranty service requests you would receive if every homeowner used your system exactly as you describe it in your homeowner manual. For example, if you build 30 homes a year and your warranty procedures include a 60-day visit and a reminder letter at 11 months, that system would generate 60 lists per year. Adjust this by 20 percent for out of warranty, emergency, or between standard checkpoint lists and you end up with 72 or an average of 6 lists a month. Compare that to what actually comes in.

Next, considering the size and complexity of your product, how many warranty items per list do you believe to be reasonable? Take that number times 72 lists per year. Compare that total with the number of items your homeowners are actually reporting. If you believe eight items per list is a reasonable number that would total 576 items per year. If history shows that you handle 984 items per year, it's time to look at what factors are causing so much work and expense. Begin by looking at the nature of the items going through the system.

Nature of Warranty Items

Builders often analyze repair work based on the number of outstanding lists. "We have 21 outstanding lists." Interesting, but insufficient. You cannot tell how much work those lists represent or how much of it has been completed. Nor can you tell anything about the nature of the work needed. Structural? Cosmetic? Material failure? Workmanship? Natural causes? On the other hand, when a builder says, "We have 108 work orders containing 417 items; 9 percent are cabinet problems, 11 percent are floor covering . . ." a clearer picture comes into focus.

Completion of Warranty Items

Accurate information on completion requires feedback from the trades. Return of the signed work order is the best method because this also provides documentation for the file. With typical computer supported systems, a report can be produced showing the homeowner, the service person or company, items to be corrected, and due date. Warranty personnel make notes on the printed reports and usually on a weekly basis the administrator updates the information.

This same data can be sorted by trade. Consistent failure of one or two trades to meet your required response time must result in some action on the part of the company. Failure to do this sends a dangerous message. "We know these trades are slow but we are letting them get away with it." On the other hand, when you replace a trade whose poor performance has failed to improve after several months and as many conversations, this sends an equally clear message.

Warranty Budget

For years builders estimated warranty expense at .5 percent of the sales price of the home ($500 for a $100,000 home). Some builders set aside as much two percent for warranty service; others (usually through good quality management) have warranty expense as low as .3 of one percent of sales revenue. When warranty costs exceed what you believe to be reasonable amounts for your company, look beneath the surface to discover the causes. Consider:

- Poor design or purchasing decisions
- Inadequate supervision during construction, lack of a quality control program, or work overload
- Delivery of incomplete homes
- Failure to define warranty standards for buyers
- Lack of training of warranty reps
- Paying trades for warranty work or failure to back charge
- Inadequate warranty staff
- Coding errors
- Denied items

Who's Responsible?

Another raging debate in the home building industry, similar to the question of who should conduct orientations, is who should oversee warranty work. The right way to staff and organize warranty service is the way that is appropriate to your circumstances and allows you to meet your customer satisfaction goals. To find the best method for your company, consider three activities: management, administrative, and technical.

For small volume companies, having a warranty rep who also carries tools to perform repairs often works well. This person may need administrative support in the office. In another arrangement, one person inspects items, performs administrative tasks, and is supported by a technician (who may also work part time for construction doing punch out items). In both approaches, trade contractors become involved as needed.

Some companies believe that having superintendents manage warranty work on their sites creates an incentive for them to build a better home. Following this logic, builders should have the superintendents perform accounting to give them an incentive to stay within the house budget. Whatever lessons may be learned, having superintendents oversee warranty distracts them from their primary task: building quality homes. Further, this procedure creates opportunities for construction to bury the bodies, effectively eliminating any quality management check and balance. And in the conflict of interest that results warranty generally loses.

When production reaches an appropriate number–usually around 40 to 50 homes per year–one person should manage warranty service and report directly to the company's executive officer, president, or owner. Having the warranty manager report to the head of construction or sales leads to another conflict of interest, in which–again–warranty generally loses.

In-house Warranty Technicians: How Many Are Enough?

Calculating how many warranty technicians a company needs is a simple matter of dividing the number of homes to be built by the number of homes one technician can care for. A commonly accepted ratio suggests one warranty technician for every 75 homes under warranty. Unfortunately this question cannot be answered that simply. Many factors influence this decision.

Geography. As driving time increases, the number of homes one person can service decreases.

Quality Control. Strong quality control means fewer warranty items per house, so one technician can service more homes.

Size and Design of Product. Smaller homes usually have fewer warranty items than larger homes, even at the same quality level. Design contributes as well. A small home with a lot of sophisticated details can generate more warranty repairs than a large plain home.

Division of Work. Which repairs are done by in-house staff and which by trade contractors?

Courtesy Repairs. Although the investment in courtesy repairs is often rewarded by referrals, this commitment does affect workload for warranty technicians.

Competitor's Services. If a competitor announces some terrific new service, pressure builds to respond in kind.

Experience and Training. A warranty service veteran can complete work orders faster and therefore may be well worth the extra dollars.

Materials and Parts. Having materials or equipment readily available increases efficiency and the number of homes one person can service.

Customer Service History. Satisfied customers make more reasonable service requests.

Expectations. If company promises exaggerate, more warranty technicians will be needed.

Sales Fluctuations. Unless you have implemented even flow production (start the same number of homes each week), the number of service technicians required to meet company goals may change with fluctuations in sales. Suddenly what began as a simple formula becomes quite complicated. Working backward from your service goals may be an easier approach. Does repair quality meet company standards? Are repairs completed within the desired time limits? Is documentation adequate? Do other departments receive useful feedback from warranty? Are homeowners satisfied? Are costs reasonable? If the answer to any of these questions is no, analysis is in order. You may need more people, different people, more training, or better technical support.

Administrative Staff

Administrative staffing levels are affected by nearly all of the same factors as technical staff. If results fail to meet company goals, careful study is needed. Eliminate inefficiencies and increase training among current staff before adding more people. If computer support is in place, is it used? Administrators sometimes continue to use the old manual system as a comfortable back-up, doubling their work load. Adequate training develops confidence in the new system. If you confirm that current staff is efficient but unable to keep up, temporary help or additional permanent staff may be the answer. If you have not computerized this function, doing so may be a more economical long-term solution.

Centralized or Decentralized Warranty Service

Another question that industry veterans vigorously debate is which works best-centralized or decentralized warranty service? In truth, neither is intrinsically "best." Your search is for the method that is most appropriate to your situation. What works well this year may need to be adjusted next year. Consider such factors as those listed below.

Geography. Distance is the most obvious aspect of this, but even short distances in heavy traffic can be detrimental to service. Tolls and road conditions may enter into your thinking, as well as where your employees live relative to the work site(s).

Control. Will management be comfortable without daily observation of the activities of field service personnel? If you trust that people you have hired have the integrity to put in a full day's work and the knowledge to do so effectively this should not be an issue.

Physical Facilities. Is there a suitable location for an on-site warranty office? A desk, phone, fax, possibly computer, files, and so on are important support tools.

Documentation. How will documentation be managed? Consistency can quickly become the victim of decentralizing unless clear procedures are in place and all personnel follow them.

Personnel. Are sufficient numbers of skilled personnel available to staff separate warranty offices? How will those personnel stay connected to the rest of the company: weekly in-office staff meeting? Email, phone? Will the department head make frequent field visits?

Close Out. How long will you need the satellite office? Does the time frame make sense when balanced against the set up cost?

Outsourcing

Outsourcing companies offer a variety of services, ranging from administrative to physical repairs. The choices might include inspections during construction, delivery preparation, orientation, and warranty services (inspections, work orders, repairs, follow up, and record keeping). Some builders address their warranty obligations with an outside company that specializes in warranty service administration. If you are considering this, keep these points in mind:

- Study the procedures homeowners would be expected to follow. Are they reasonable-from a homeowner's point of view? Do the procedures support the image you want?
- Will the outsourcing company work from your standards or do they insist on a one-size-fits-all approach based on so-called "industry standards"?
- Some outsourcing firms charge a flat fee, usually based on a percentage of the sales price of the home. Others charge an hourly fee, and still others use a combination of the two.
- Is the literature buyers receive attractive, well-written, and useful? Are grammar, spelling, and punctuation correct? Is the information well-organized?
- The primary goals of the outsourcing company should be to produce results for your homeowners and, by doing so, to enhance your reputation. Are response times and tracking systems aggressive enough to accomplish these goals?
- Ask to meet the personnel who would be assigned to your communities. Are these people you would hire? How much turnover occurs?
- What documentation and reports will the warranty service company provide and when? Will the information help you identify recurring items or defend your company in a legal action?

What's in a Name?

Wanting to create a friendly image, many builders have named their warranty office the "customer care" department. Ironically, this can result in lowering buyer satisfaction rather than raising it.

Big Picture Implications

The old adage "Customer service is an attitude, not a department" applies. Having a department named "customer care" (or for that matter, "customer service") implies to other employees that customer service is the responsibility of the "customer care department" when in fact customer service is part of every job.

Homeowner Expectations

Builders work hard to align customer expectations. What a shame to undo a large part of that effort by naming warranty "customer care." Customer care implies flexibility and a wider range of attentions. What follows in most cases is warranty attention based on defined warranty standards, which contrasts sharply with the implications of the friendly name. The title "New Home Warranty Department" on the other hand, indicates that specific repairs are available during a specific time period. However, nothing in this name prohibits a builder's warranty office from considering unusual circumstances and making common sense exceptions when appropriate.

Survey Savvy

Perhaps the least apparent but nonetheless a significant concern to be aware of is the confusion that the department names "customer care" or "customer service" can create with customer satisfaction surveys. When homeowner surveys ask questions about "service" the person or company creating the questions has certain performance areas in mind-usually warranty. The problem comes from the assumption that the homeowners' answers apply to the same performance areas. Customers see a company's service as fluid, coming from all personnel and all directions, flowing in an around the transaction from start to finish. So what "service" is the customer rating: phone calls not returned by sales? pricing slow to come from design? Or warranty service? The name New Home Warranty Department reduces confusion.

The complexity of warranty service combined with homeowner's subjective evaluation of quality, create a challenge for builders. Meeting that challenge requires planning and close attention to detail. Once an effective system is in place, constant monitoring is essential to keep it effective. The effort is repaid with satisfied customers and all the benefits they can generate.

Conclusion

One final challenge and caution: Beware of your own success. Outstanding service achievements have lulled many once successful organizations into complacency. Intrinsic in this complacency are the dangers of listening less carefully (we already know everything), improving more casually (it's not broke, so what's there to fix?), and communicating arrogance instead of appreciation (we're so good, if one customer is unhappy, so what?).

Avoid such complacency by expecting your service goals and methods to evolve. To maintain a dynamic service program, invest the time necessary to stay in touch with your frontline personnel and your customers. Gather and study feedback from all sources and watch the horizon for meaningful trends. Just as you revise and update floor plans, your plans for earning the respect and loyalty of customers will also evolve. Work for your customers' satisfaction continuously, just as you strive for product innovation, quality, and control of costs. Customer satisfaction is the result of good habits, constant vigilance, and willingness to make appropriate changes as conditions evolve.

This evolutionary process generates change. Many find change uncomfortable; it brings questions, surprises, and work. Resistance is a normal stage of coping with change. Neither service programs nor reputations improve quickly. Old habits change gradually; new habits become comfortable over time, not over night. A year's worth of homeowners are in the service pipeline with the old problems built into the company's relationship with them.

But one day someone notices the atmosphere is friendlier. Angry calls and threatening letters have diminished. Once improvement begins, it continues, gradually picking up momentum. It takes time, but it is worth it. Just ask a customer.

Resources

A book of this length is necessarily limited to insights, highlights, and essentials. Whether you are reading it to refine an existing approach to service or to develop one for a new company, you may encounter subjects where you need more detail. These resources offer more detailed information about many of the subject covered here.

Fredley, John and John Schaufelberger. *Contracts with the Trades: Scope of Work Models for Home Builders.* Washington, DC: BuilderBooks, National Association of Home Builders, 1997, 124 pp.

Haasl, John J. and Peter Kuchinchy II. *Production Checklist for Builders and Superintendents.* Washington DC: BuilderBooks, National Association of Home Builders, 2000, 139 pp.

NAHB Business Management. *Building Quality: An Operations Manual for Home Builders.* Washington DC: BuilderBooks, National Association of Home Builders, 2000.

Jaffee, David S. and David Crump. *Contracts and Liability,* 4th edition. Washington, DC: BuilderBooks, National Association of Home Builders, 2004, 143 pp.

Jaffee, David S. *Warranties and Disclaimers for Builders,* 4th ed. Washington, DC: BuilderBooks, National Association of Home Builders, 1999, 72 pp.

Kohn, Alfie. *Punished by Rewards.* Boston: Houghton Mifflin Company, 1993, 398 pp.

Nelson, Bob. *1001 Ways to Reward Employees.* New York: Workman Publishing. 1994, 275 pp.

Smith, Carol. *Customer Relations Handbook for Builders.* Washington, DC: BuilderBooks, National Association of Home Builders, 1998, 435 pp.

Smith, Carol. *Dear Homeowner.* Washington, DC: BuilderBooks, National Association of Home Builders, 2000, 160 pp.

Smith, Carol. *Homeowner Manual,* 2nd edition. Washington, DC: BuilderBooks, National Association of Home Builders, 2001, 208 pp.

Smith, Carol. *Meetings with Clients.* Washington DC: BuilderBooks, National Association of Home Builders, 2001, 160 pp.

Smith, Carol. *Warranty Service for Home Builders.* Washington DC: Builder Books, National Association of Home Builders, 2002, 146 pp.

For more information about products from BulderBooks, visit www.BuilderBooks.com or call 800-223-2665. To reach Carol Smith visit *www.cjsmithhomeaddress.com* or call 719-481-6247.

Index

Page numbers in *italics* refer to illustrations